Library of
Davidson College

STUDIES IN
TROPICAL AMERICAN MOLLUSKS

STUDIES IN TROPICAL AMERICAN MOLLUSKS

Edited by

FREDERICK M. BAYER

AND

GILBERT L. VOSS

UNIVERSITY OF MIAMI PRESS

Coral Gables, Florida

Editorial Committee for this volume:

GILBERT L. VOSS
FREDERICK M. BAYER
ROBERT AUSTIN SMITH

Copyright © 1971 by
University of Miami Press

Library of Congress Catalog Card Number 70-170142
ISBN 0-87024-230-X

All rights reserved, including rights of reproduction and use in any form or by any means, including the making of copies by any photo process, or by any electronic or mechanical device, printed or written or oral, or recording for sound or visual reproduction or for use in any knowledge or retrieval system or device, unless permission in writing is obtained from the copyright proprietors.

Manufactured in the United States of America

Contents

Foreword	vii
Cephalopods Collected in the Gulf of Panama *By Gilbert L. Voss*	1
Mollusks from the Gulf of Panama *By Axel A. Olsson*	35
The Conidae of the PILLSBURY Expedition *By James Nybakken*	93
New and Unusual Mollusks Collected *By Frederick M. Bayer*	111

Foreword

This volume consists of four studies on tropical American marine mollusks obtained by expeditions of the Rosentiel School of Marine and Atmospheric Science of the University of Miami. They were published originally in series in the quarterly *Bulletin of Marine Science* and are here made available in a separate volume to all those who are interested in the rich molluscan fauna of this region and to whom the regular issues of the *Bulletin* may not be so readily accessible.

These papers form part of the results of the Deep-Sea Biology Program of the University of Miami. This program arose because of the dearth of reliable information on the nature and distribution of the western Atlantic molluscan fauna. This program was conceived as a systematic survey of the marine biota of the vicinity of the Straits of Florida and was commenced in 1962 aboard the school's research vessel, GERDA, and the school's larger research vessel, the JOHN ELLIOTT PILLSBURY, which made it possible to extend this survey to broader tropical Atlantic region exclusive of the Gulf of Mexico and to the crucially important faunal areas to the east and west—the Gulf of Guinea and the Bay of Panama—which contribute so heavily to the present faunal characteristics of the Caribbean area.

Because knowledge of these regions depends almost entirely upon scattered collections made during biological expeditions many years ago, we have undertaken a sampling program designed to explore the region systematically, using several types of standard sampling gear operated in uniform manner. The research collections have been processed in the same way and so far as possible are being analyzed by the same corps of specialists within and outside of the School of Marine and Atmospheric Science.

This program could never have been commenced without the help of the National Science Foundation through grants for operation of vessels for biological research (GB-1204 and GB-7082), and the scientific results could not have been obtained without the support of the Foundation to individual scientists who have studied the collections. The continuing systematic field program could never have attained its comprehensive coverage without the farsighted and generous aid of the National Geographic Society through its grant for the study of deep-sea biology, which also has supported many aspects of the laboratory research and analysis of the rich collections that have been obtained. These collections are so large and so comprehensive that their study and final interpretation will require several additional years of concentrated research. So much new information has come to light that it is desirable to make as much of it as possible available to researchers before the total studies are published. It is with this objective in mind that the four papers combined in this volume have been published.

All of these papers deal more or less exclusively with the faunas on either side of the Isthmus of Panama and are the outgrowth of studies undertaken in connection with the feasibility and impact of the proposed interoceanic sea-level canal. All these studies are a direct contribution to an accumulation of base-line knowledge that will be of critical importance if a new canal is opened. This will be no less, even if such a canal never is constructed. The Caribbean and Pacific coasts of Middle America of the present day are descended from what was once a broad tropical American faunal province, and the thorough knowledge of forms as they exist will provide a deeper insight into their development through geologic ages. Thus, if man can alter the marine environment on such a scale as by the artificial reuniting of the Atlantic and Pacific faunas, he inevitably will continue to do so in countless lesser ways, and studies such as these are needed to enhance our knowledge of the marine biota before the environmental changes proceed further.

THE EDITORS

Biological Results of the University of Miami Deep-Sea Expeditions. 76.
CEPHALOPODS COLLECTED BY THE R/V JOHN ELLIOTT PILLSBURY IN THE GULF OF PANAMA IN 1967[1]

GILBERT L. VOSS
University of Miami, Rosenstiel School of Marine and Atmospheric Sciences

Abstract

The cephalopod fauna of the Gulf of Panamá is reviewed, based mainly upon collections made by the R/V JOHN ELLIOTT PILLSBURY during 1967. Records are given of 812 specimens, of which four species and one genus are new to science: *Octopus selene, O. balboai, O. stictochrus* and *Euaxoctopus panamensis*. Twenty-six species are now known from the Gulf, of which 11 are new records first reported upon here. The relationships of the fauna of the Gulf of Panamá with those of the eastern Pacific and southwestern Caribbean Sea are discussed.

Introduction

The projected plans for constructing an interoceanic sea-level canal across the Isthmus of Panama to connect the waters of the Gulf of Panamá and the Caribbean Sea have aroused international interest in the faunas of the two regions. The fauna and flora of both areas are very poorly known, and before the land barrier is removed it is essential that studies be undertaken to list and describe the life involved. As part of the Rosenstiel School of Marine and Atmospheric Sciences' studies of this problem, the R/V JOHN ELLIOTT PILLSBURY entered the Gulf of Panamá in May 1967, to occupy a series of planned trawl stations on the continental slope and adjacent waters. The research conducted was part of a long-term study of the shallow- and deep-sea fauna of the tropical Atlantic and related regions. The research program was supported by the National Geographic Society–University of Miami Deep-Sea Biology Program. The ship time was supported by National Science Foundation grant GB-5776. The work on the octopods was subsidiary to National Science Foundation grants GB-1090 and GB-5729X.

---

[1] Contribution No. 1309 from the University of Miami, Rosenstiel School of Marine and Atmospheric Sciences. This paper is one of a series resulting from the National Geographic Society–University of Miami Deep-Sea Biology Program. The work was supported in part by National Science Foundation grants GB-5776, GB-1090, and GB-5729X, and the National Geographic Society–University of Miami Deep-Sea Biology Program.

During the course of the cruise, 92 stations were occupied within the Gulf of Panamá from the shore to over 3200 meters. An account of the cruise and the station data have been given by the writer (Voss, 1967). Eight hundred and twelve specimens of cephalopods were taken, representing 17 species, of which four species and one genus were new to science.

I wish to thank Dr. Jon Staiger for his generous assistance in the sometimes arduous task of conducting the cruise. Without his help, little time would have been available to me for conducting shipboard studies of the cephalopods. I also wish to thank Dr. Lipke Holthuis and the graduate students of the School of Marine and Atmospheric Sciences who participated in the cruise, made the collections and contributed so greatly to its success. In particular I wish to thank Dr. Richard Young and Dr. Edward McSweeny, my graduate student colleagues in cephalopod research, who were largely responsible for the care, handling, and preserving of cephalopods collected and who made many observations on behavior and color. I wish to thank Dr. C. F. E. Roper and the Smithsonian Institution for the loan of a small collection of octopods from Panamá, including the type of *Octopus balboai*, and C. E. Dawson of the Gulf Coast Research Laboratory for permission to study a small collection of octopods from the coasts of Panamá and Costa Rica. The measurements and indices given in this paper are those defined by Voss (1963). The illustrations were executed by Constance Stolen McSweeny, to whom grateful thanks are extended.

The synonymies given in this paper are restricted to records of specimens from the Gulf of Panamá or to discussions concerning these records. The literature cited is likewise restricted to those papers dealing directly with the area under consideration.

Historical Résumé

The cephalopod fauna of the Gulf of Panamá is very poorly known. Only three major expeditions have made even the most fragmentary collections from its waters and incidental shore collections have been inconsequential. Even some of the latter are not known to have been made specifically in the Gulf of Panamá. The present review is not exhaustive but it is believed to be nearly complete.

The first report of a cephalopod, presumably from the Gulf of Panamá, was given by Verrill (1883: 122), in his description of *Octopus bimaculatus*. He stated: "Numerous small specimens were obtained at Panamá and on the coast of San Salvador by Mr. Frank H. Bradley, for the Museum of Yale College, in 1866 and 1867." Pickford & McConnaughey (1949) considered these specimens to represent an as yet undescribed species, not a member of the *bimaculatus* complex.

In 1889, Jatta described the cephalopods taken on the circumnavigational cruise of the Italian school ship VETTOR PISANI. Among these was an unusual barred octopus, *Octopus chierchiae,* from the coast of Panamá, presumably in the gulf. This species was not again seen until Voss (1968) described and illustrated it anew from the PILLSBURY cruise of 1967.

The original fisheries research vessel ALBATROSS worked the entrance to the Gulf of Panamá in 1891, under the direction of Alexander Agassiz. No stations were made directly on the continental shelf; they were confined to the slope and the deep waters southward of the parallel of Punta Mala. For the purpose of this report, stations west of 81°W and south of 6°45′N are not included. Eight species of cephalopods were reported upon from this area by Hoyle (1904): *Calliteuthis reversa, Polypus* sp., *Tremoctopus scalenus, Bathyteuthis abyssicola, Moschites verrucosa, Mastigoteuthis dentata, Abraliopsis hoylei,* and *Abraliopsis* sp. Most of these specimens were taken at the deeper stations near the outer limits proposed here.

S. Stillman Berry, in 1911, described a new species of *Lolliguncula, L. panamensis,* from "Panama." It is known to occur in the Gulf of Panamá. In his monographic study of the oegopsid cephalopods, Pfeffer (1912) listed only two species from the area: *Abralia (Nepioteuthion) panamensis* (=*Abraliopsis* sp. Hoyle, 1904) and *Abralia (Micrabralia) affinis* (=*Abraliopsis hoylei,* Hoyle, 1904). Both of these species, however, belong to the genus *Abraliopsis* as originally cited by Hoyle (1904). *A. panamensis* was based upon a larval specimen and is here considered a *species dubia.*

Robson (1929, 1932), in his monograph of the Octopodidae, listed 12 species of octopods from "Panama." Most of these were citations of species given above or species from outside our area of reference. It would serve no purpose to relist them here, except for *Argonauta argo, A. nouryi,* and *A. cornutus,* all without specific locality data.

In 1921, Schmidt took the DANA briefly into the eastern Pacific. The material from this cruise was reported upon by Joubin (1929, 1931). Among the new species described were three emanating from the Gulf of Panamá or its immediate environs: *Retroteuthis pacifica, Valbyteuthis danae* and *Drechselia danae,* the latter just south of latitude 6°45′N, at 6°40′N.

Roper (1968) described a new species of *Bathyteuthis, B. bacidifera,* from the Eastern Pacific Equatorial Water Mass. A specimen collected on the PILLSBURY cruise served as one of the paratypes, as well as a specimen taken by the DANA from the environs of the Gulf of Panamá. In 1969, Roper recorded additional DANA material belonging to this species from the same area.

In 1968, I redescribed and illustrated *Octopus chierchiae* from stations deep within the gulf. No additional specific records of cephalopods from this region are known to me. However, a few species should be included, because of their wide range along the west American coast. These are *Dosidicus gigas* and *Symplectoteuthis oualaniensis.* The first is confirmed in the present report; surprisingly, no specimens of the latter were taken during the PILLSBURY cruise.

A list of the species now known to occur within the Gulf of Panamá is given below. Those whose names are preceded by single asterisks were taken also by the PILLSBURY. The names given are those presently accepted by specialists in this group. The names given in prior reports are enclosed within parentheses preceded by an = sign. Double asterisks represent new records for the Gulf of Panamá.

LIST OF CEPHALOPODS KNOWN TO OCCUR IN THE GULF OF PANAMÁ

***Loliolopsis diomedeae* (Hoyle, 1904), new combination
**Lolliguncula panamensis* Berry, 1911
**Abraliopsis affinis* (Pfeffer, 1912) (=*Abraliopsis hoylei,* Hoyle, 1904)
***Pterygioteuthis giardi* (?) Fischer, 1895
***Onychoteuthis banksi* (Leach, 1817)
**Bathyteuthis bacidifera* Roper, 1968
Bathyteuthis abyssicola Hoyle, 1885
**Dosidicus gigas (d'Orbigny, 1835)
Symplectoteuthis oualaniensis (Lesson, 1830)
Histioteuthis sp. (=*Calliteuthis reversa* Verrill, 1880)
Mastigoteuthis dentata Hoyle, 1904
Valbyteuthis danae Joubin, 1931
Drechselia danae Joubin, 1931
***Japetella diaphana* Hoyle, 1885
***Octopus selene,* new species
***Octopus balboai,* new species
***Octopus stictochrus,* new species
**Octopus chierchiae* Jatta, 1889
** ? *Octopus oculifer* (Hoyle, 1904) (?=*O. bimaculatus* Verrill, 1883, *pars*)
** ? *Octopus vulgaris* Cuvier, 1797
***Euaxoctopus panamensis,* new genus & new species
**Argonauta pacificus* Dall, 1872
**Argonauta cornutus* Conrad, 1854
*Argonauta nouryi Lorois, 1852
Tremoctopus violaceus delle Chiaje, 1830 (=*Tremoctopus scalenus* Hoyle, 1904)
Vampyroteuthis infernalis Chun, 1903 (=*Retroteuthis pacifica* Joubin, 1929)

GEOGRAPHICAL DISCUSSION

Any detailed discussion of the geographical distribution of cephalopods is premature. There are still many areas that have not been sampled and reported upon. In addition, newer types of trawls are capturing species new to old and well-worked areas. Nevertheless, it is possible to show certain broad affinities and patterns, always realizing that these are subject to modification with addition of new data.

The picture of cephalopod distributional patterns in the eastern Pacific is complicated by our general lack of knowledge of the cephalopod faunas along the western coasts of North and South America. Despite the long history of cephalopod research in California, primarily the work of S. Stillman Berry, there is no modern monographic treatment of the cephalopods of those waters. The situation with regard to the octopods is particularly frustrating. Numerous species of small octopods have been described, but the descriptions are very brief, there is a complete void of morphometric data, and most have never been illustrated. The situation with regard to western South America is even worse; here we must rely almost solely upon the mid-nineteenth century work of Alcide d'Orbigny.

The presently known species of cephalopods recorded from the Gulf of Panamá and its environs amount to 26 species. An analysis of these shows that they are divisible into four faunal components: Endemic Tropical Eastern Pacific, Temperate Eastern Pacific, Indo-Pacific, and Cosmopolitan Warm-Temperate.

TABLE 1
LIST OF ENDEMIC TROPICAL EASTERN PACIFIC CEPHALOPODS

Loliolopsis diomedeae	*Octopus selene*
Lolliguncula panamensis	*O. balboai*
Abraliopsis affinis	*O. stictochrus*
Drechselia danae	*O. chierchiae*
Mastigoteuthis dentata	*O. oculifer*
	Euaxoctopus panamensis

Endemic Tropical Eastern Pacific.—This is the largest faunal element and amounts to 42 per cent of the total, or 11 species. In contrast to the Gulf of Guinea (Voss, in press) a somewhat comparable area that has no endemic genera, this region possesses three—*Loliolopsis, Drechselia,* and *Euaxoctopus. Loliolopsis* and *Euaxoctopus* will probably prove to be truly endemic; *Drechselia* is doubtful, because it is a bathypelagic form, and such forms usually have wide oceanic distributions. *Loliolopsis diomedeae* is unique. *Lolliguncula panamensis* is obviously a Tethyan species and is closely related to *L. brevis* from the western Atlantic, with which it may be identical. The genus is represented in the eastern Atlantic by *L. mercatoris,* and in the western Indian Ocean by *L. abulati. Abraliopsis affinis* is closely related to an undescribed species in the Atlantic and other species in the tropical Pacific and Indian oceans. *Mastigoteuthis dentata,* similar to *Drechselia danae,* is bathypelagic, but so far is known only from the Gulf of Panamá and the Galapagos Islands.

The octopods—*Octopus selene, O. balboai, O. chierchiae, O. oculifer,* and *Euaxoctopus panamensis*—are all endemic. *O. chierchiae* is closely related to *O. zonatus* from the Caribbean, and *O. oculifer* may be a Pacific counterpart of *O. hummelincki* from the Caribbean. *Euaxoctopus* has no known representative elsewhere.

TABLE 2
LIST OF TEMPERATE EASTERN PACIFIC CEPHALOPODS

Dosidicus gigas	*Argonauta pacificus*

Temperate Eastern Pacific.—These cephalopods are found well beyond the boundaries of the tropical eastern Pacific region, occurring southward perhaps as far as Chiloe Island and northward off California. There are only two species, which account for 8 per cent of the fauna.

TABLE 3
LIST OF INDO-PACIFIC CEPHALOPODS

Symplectoteuthis oualaniensis	*Argonauta cornutus* *A. nouryi*

Indo-Pacific.—It is surprising that so few cephalopods from the great Indo-Pacific area have found their way to the coasts of the Americas. These would of necessity, in order to cross the eastern Pacific basin, be either oceanic species or forms with long-duration planktonic larvae. Only three are so far known, accounting for 12 per cent of the fauna. *Symplectoteuthis oualaniensis* is an Indo-Pacific oceanic and slope-associated form. *Argonauta cornutus* and *A. nouryi* are somewhat problematical species which appear to have a wide distribution in the Central Pacific area.

TABLE 4
LIST OF COSMOPOLITAN WARM-TEMPERATE CEPHALOPODS

Pterygioteuthis giardi *Onychoteuthis banksi* *Bathyteuthis abyssicola* *B. bacidifera* *Histioteuthis* sp. *Valbyteuthis danae*	*Japetella diaphana* *Octopus vulgaris* *Tremoctopus violaceus* *Vampyroteuthis infernalis*

Cosmopolitan Warm-Temperate Cephalopods.—This is the second largest group, comprising 38 per cent of the fauna of the Gulf of Panamá. The decapods all belong to the bathypelagic zone, an area noted for the wide distribution of its inhabitants. The octopods are either pelagic (*Japetella diaphana* and *Tremoctopus violaceus*) or bathypelagic (*Vampyroteuthis infernalis*). *Octopus vulgaris* lays small eggs and has small, planktonic larvae that can maintain themselves in the plankton for long periods of time. This species is the most widely distributed of all benthic octopods.

Among the cephalopods of the Gulf of Panamá are three species which have perhaps their closest relatives in the Caribbean Sea. These have often been termed geminate species. One of the most striking of these is *Octopus chierchiae*. Its relative in the Caribbean is *O. zonatus*, from which it is quite distinct but can be distinguished only by careful examination. Undoubtedly these two were derived from a common immediate ancestor, or one from the other. Similarly, *Octopus oculifer* is very closely related to *O. hummelincki*, the common ocellated octopus of the Caribbean. *O. oculifer* is also, however, closely allied perhaps to the *bimaculatus-bimaculoides* complex of the California coast. In the case of the *Lolliguncula panamensis–L. brevis* complex, we are, as stated above, dealing with a Tethyan distribution. However, so few specimens of *L. panamensis* have been taken that its status is still somewhat problematical.

A comparison of the faunas of the total tropical eastern Pacific and the Caribbean will somewhat change the picture presented here. As our sampling, however, was restricted to the Gulf of Panamá and records outside of it are few and doubtful, the larger analysis must be delayed until better data are available.

Order TEUTHOIDEA

Suborder Myopsida

Family Loliginidae

Loliolopsis diomedeae (Hoyle, 1904), new combination

Loligo diomedeae Hoyle, 1904: 29, pl. 5, fig. 13; pl. 6, figs. 1-7.
Loliolopsis chiroctes Berry, 1929: 267, figs. 1-9, pls. 32-33.

Material Examined.—Type of *Loligo diomedeae*, ♀, mantle length 85 mm, ALBATROSS Sta. 3422, off Acapulco, 16°47′30″N, 99°59′30″W, 141 fm, green mud, April 12, 1891, USNM 574847.—1 ♂, mantle length 79.0 mm, PILLSBURY Sta. 531, 8°25.5′N, 79°10.1′W, west of the Perlas Islands, Panamá, in 57-64 meters with 10-ft OT, May 6, 1967.—2 ♂, mantle length 66-58 mm, PILLSBURY Sta. 548, 8°09.8′N, 78°25′W, in 18-20 meters off Punta Garachiné, Panamá, with 16-ft OT, May 7, 1967.—2 ♂, mantle length 49-59 mm, PILLSBURY Sta. 546, 8°19.2′N, 78°35.8′W, in 27-31 meters in the Bahia San Miguel with 10-ft OT, May 7, 1967.—2 ♂, mantle length 44-46 mm, PILLSBURY Sta. P-495, 7°59.2′N, 80°00.2′W, in 40-37 meters off Punta Mala, Panamá, with 10-ft OT, May 2, 1967.—3 ♀, mantle length 99-106 mm, PILLSBURY Sta. 563, 8°14.8′N, 78°56.2′W, at the surface with night light and dip net off Punta del Concholon, Isla del Rey, Perlas Islands, May 8, 1967.—4 ♂, mantle length 57-78 mm, 5 ♀, mantle length 89-102 mm, PILLSBURY Sta. 534, 8°54.5′N, 79°10.2′W, in 17 meters off Isla Chipillo with 10-ft OT, May 6, 1967.—176 ♂, mantle length 60-92 mm, 247 ♀, mantle length 63-97 mm, PILLSBURY Sta. 518, 8°00′N, 79°31.1′W, in 99-95 meters in the middle of the Gulf of Panamá with 40-ft

OT, May 4, 1967.—8 ♂, mantle length 64-77 mm, 156 ♀, mantle length 88-113 mm, PILLSBURY Sta. 554, 7°59′N, 79°02.2′W, at surface south of the Perlas Islands with night light and dip net, May 7, 1967.

Remarks.—This species occurs in large numbers in the Gulf of Panamá and should be able to support a considerable fishery. The females seem to be more numerous than the males (about a 2 : 1 ratio), but collecting methods were not adequate for eliminating bias. The males are smaller than the females and are easily distinguished by the great length and whiplike appearance of the left ventral arm. The right ventral arm is normal, except for the development of a broad ventral membrane about midway of the arm.

When the schools were drawn together under the night light on several occasions, the animals went into a mating frenzy, dashing about frantically and discharging spermatophores at random.

The animals conform well with the description given by Berry (1929). The species is known from the Gulf of California to Peru.

Discussion.—In 1904, Hoyle reported upon the collections made by the ALBATROSS off Central America and the west coast of Mexico. In his report, he described a new species of *Loligo, L. diomedeae*, based upon a single female from off Acapulco. No additional information has been given for this species since Hoyle's report. The type is in the U. S. National Museum and is now so hardened that little can be determined from examination of it.

In 1929, Berry described a new genus and species, *Loliolopsis chiroctes*, from the Gulf of California. The male is distinctive, and Berry erected a new genus to contain it. It was well illustrated and described.

During the PILLSBURY cruise of 1967 in the Gulf of Panamá, females of a loliginid were collected under the night light and, when checked against Hoyle's paper, were immediately identified as *Loligo diomedeae*. However, when males were examined, they were found to conform to Berry's description of *Loliolopsis chiroctes*. A careful check of the two descriptions and an examination of the type of *L. diomedeae* have convinced me that the two species are identical. As Hoyle's name has precedence by some 23 years, Berry's name *chiroctes* must unfortunately be placed in synonymy. His generic name, however, is apparently valid, and the new combination results in the name *Loliolopsis diomedeae*.

Lolliguncula panamensis Berry, 1911

Lolliguncula (?) *panamensis* Berry, 1911: 100, figs. 1-7, pl. 6.

Material Examined.—2 ♀, mantle length 46-52 mm, PILLSBURY Sta. 537, 8°35.7′N, 78°41.7′W, at surface between the Perlas Islands and Río Hondo, Panamá, with night light and dip net, May 7, 1967.

Only two small females were taken. They seem assignable to this species, which is known to occur from Panamá to Ecuador. Much more needs to be learned about this poorly described and little studied species.

UNIDENTIFIED LOLIGINIDS

Material Examined.—1 juvenile, mantle length 18 mm, from PILLSBURY Sta. 548.—1 juvenile, mantle length 21 mm, PILLSBURY Sta. 546.—1 juvenile, mantle length 12 mm, PILLSBURY Sta. 534.—4 juveniles, mantle length 20.0-35.0 mm, PILLSBURY Sta. 544.

The above listed juveniles are too immature for positive identification, due to our lack of knowledge of the life histories of Panamanian loliginids.

LOLIGINID EGG MASSES

Large egg masses of the type deposited by loliginids were collected at the following PILLSBURY stations: 488, 534, 544, 547, 548. Since individuals of *Loliolopsis diomedeae* were spawning at this time, it is possible that these eggs can be associated with this species. Individual egg capsules were up to 35 mm long and contained numerous eggs enclosed in the gelatinous mass.

Suborder Oegopsida

Family Enoploteuthidae

Abraliopsis affinis (Pfeffer, 1912)

Abraliopsis hoylei, Hoyle, 1904: 36, pl. 1, fig. 3; pl. 8, pl. 10, figs. 1-10 (not *Abraliopsis hoylei* [Pfeffer, 1884]).
Abralia (Microbralia) affinis Pfeffer, 1912: 160.

Material Examined.—1 ♂, mantle length 25 mm, 5 ♀, mantle length 18-27 mm, 1 head, PILLSBURY Sta. 510, 6°54'N, 79°57.6'W, in 3182-3164 meters south of Punta Mala, Panamá, with 40-ft OT, May 3, 1967.

This is a very common species occurring from the Gulf of Panamá northward to southern California and in the tropical eastern Pacific.

Pterygioteuthis giardi(?) Fischer, 1895

Material Examined.—2 juveniles, mantle length 16-19 mm, PILLSBURY Sta. 510, 6°54'N, 79°57.6'W, in 3182-3164 meters south of Punta Mala, Panamá, with 40-ft OT, May 3, 1967.

Only two badly injured specimens of this widely distributed species were taken on the cruise.

Family Onychoteuthidae

Onychoteuthis banksi (Leach, 1817)

Material Examined.—2 juveniles, mantle length 23-34 mm, PILLSBURY Sta. 511, 7°16.2'N, 79°50.8'W, at surface off Punta Mala, Panamá, with night light and dip net, May 3, 1967.

Remarks.—This is a common pelagic species of nearly cosmopolitan distribution, but which may eventually be split up into a number of species.

Family Bathyteuthidae

Bathyteuthis bacidifera Roper, 1968

Bathyteuthis bacidifera Roper, 1968: 163, pls. 1-4, 7G, H; 1969: 43, pls. 6-10, 12G, H.

Material Examined.—PARATYPE: 1 ♂, mantle length 28 mm, PILLSBURY Sta. 510, 6°54'N, 79°57'W, south of Punta Mala, Panamá, 0-3182 meters, with 40-ft OT, May 3, 1967.

This specimen was reported upon by Roper (1968). It was captured together with two unidentified individuals of *Mastigoteuthis*. This species appears to be restricted to the waters of the Eastern Pacific Equatorial Water Mass, but possibly occurs in the Indian Ocean Equatorial Water Mass (Roper, 1969).

Family Ommastrephidae

Dosidicus gigas (d'Orbigny, 1835)

Material Examined.—5 juveniles, mantle length 51-114 mm, PILLSBURY Sta. 511, 7°16.2'N, 79°50.8'W, southeast of Punta Mala, Panamá, at surface with night light and dip net, May 3, 1967.—3 juveniles, mantle length 24-50 mm, PILLSBURY Sta. 503, 7°24.5'N, 79°45.2'W, east of Punta Mala, Panamá, at surface with night light and dip net, May 3, 1967.

This species was taken only at two stations near the mouth of the Gulf of Panamá. It is distributed seasonally from San Diego, California, to Chile. While it attains a large size, most individuals seen are the younger stages common in near-shore locations. It is not common in the Gulf of Panamá proper.

Family Mastigoteuthidae

Mastigoteuthis sp.

Material Examined.—2 specimens, sex indet., mantle length 49-83 mm, PILLSBURY Sta. 510, 6°54'N, 79°57.6'W, south of Punta Mala, Panamá, 0-3182 meters with 40-ft OT, May 3, 1967.

The two specimens were rather badly mangled by the trawl and may be unidentifiable. The genus *Mastigoteuthis* is sorely in need of a full revision based upon adequate material not yet available. The ALBATROSS in 1891 took a specimen of *Mastigoteuthis* within a few miles of this station, which Hoyle (1904) called *M. dentata*. Whether the PILLSBURY specimens belong to the same species cannot at present be decided.

Order OCTOPODA

Suborder Incirrata

Family Bolitaenidae

Japetella diaphana Hoyle, 1885

Japetella diaphana, Thore, 1949: 4.

Material Examined.—1 juvenile, mantle length 13 mm, PILLSBURY Sta. 510, 6°54′N, 79°57.6′W, south of Punta Mala, Panamá, 0-3182 meters with 40-ft OT, May 3, 1967.

This species has been reported previously from the Gulf of Panamá by Thore (1949). It is a cosmopolitan species.

Family Octopodidae

Octopus selene, new species

Fig. 1, a-i

Material Examined.—HOLOTYPE: ♂, mantle length 39.2 mm, PILLSBURY Sta. 501, 7°50.2′N, 79°50.5′W, northeast of Punta Mala, Panamá, in 68 meters with 10-ft OT, May 2, 1967. USNM 577617.

PARATYPES: 3 ♂, mantle length 33.1-34.3 mm, 6 ♀, mantle length 33.2-38.9 mm, PILLSBURY Sta. 498, 8°10.5′N, 79°50.2′W, western Gulf of Panamá, in 58 meters with 10-ft try net, May 2, 1967.—9 ♂, mantle length 30.0-41.0 mm, 11 ♀, mantle length 30.5-42.9 mm, PILLSBURY Sta. 500, 7°59.7′N, 79°49.7′W, western Gulf of Panamá, in 53 meters with 10-ft OT, May 2, 1967.—35 ♂, mantle length 32.2-42.8 mm, 17 ♀, mantle length 28.0-40.9 mm, PILLSBURY Sta. 501, 7°50.2′N, 79°50.5′W, northeast of Punta Mala, Panamá, in 68 meters with 10-ft OT, May 2, 1967.—1 ♀, mantle length 30.0 mm, PILLSBURY Sta. 512, 7°30.5′N, 79°41.5′W, east of Punta Mala, Panamá, in 210 meters with 10-ft OT, May 4, 1967.—2 ♀, mantle length 36.9-55.2 mm, PILLSBURY Sta. 513, 7°39.5′N, 79°40.7′W, east of Isla Iguana, Panamá, in 117 meters with 10-ft OT, May 4, 1967.—14 ♂, mantle length 29.0-40.4 mm, 21 ♀, mantle length 26.4-40.9 mm, PILLSBURY Sta. 515, 8°00.4′N, 79°40.8′W, in western Gulf of Panamá, in 79-77 meters with 10-ft OT, May 4, 1967.—1 ♂, mantle length 37.2 mm, 5 ♀, mantle length 34.7-44.1 mm, PILLSBURY Sta. 519, 7°59.4′N, 79°34.3′W, central Gulf of Panamá, in 88 meters with 10-ft OT, May 4, 1967.—4 ♂, mantle length 30.8-40.4 mm, 13 ♀, mantle length 32.1-47.9 mm, PILLSBURY Sta. 529, 8°00.7′N, 79°11.8′W, south of Isla de San Jose, Perlas Islands, Panamá, in 84 meters with 10-ft OT, May 6, 1967.—1 ♀, mantle length 58.1 mm, PILLSBURY Sta. 555, 7°50.7′N, 79°00.3′W, south of the Perlas Islands, Panamá, in 68 meters with 10-ft OT, May 7, 1967.—1 ♀, mantle length 43.2 mm, PILLSBURY Sta. 558, 7°50.7'N, 78°38.3'W, southwest of Punta Escondido, Panamá, in 62-60 meters with 10-ft OT, May 8, 1967.

Description.—This species is represented in the PILLSBURY collections by 144 specimens collected at nine stations in the Gulf of Panamá. It is a moderately small species, but very common in the places of capture. The indices given are for specimens in the 40-45 mm size-groups, although the five largest females exceeded this limit, with a maximum of 58 mm. The sexes were about evenly distributed with 69 males and 76 females.

The mantle is stout and rounded in the adult females (MWI 58.8-*7.37*-86.4), but in younger specimens is somewhat flattened. In the males (MWI 49.7-*61.9*-72.8), the mantle is more elongate and is bluntly pointed posteriorly and more flattened ventrally.

The funnel is short, stout, and is free for about half of its length. The funnel organ is W-shaped, with the outer limbs as long as the median projection. The mantle aperture is wide.

The head is narrower than the mantle and is only slightly set off by an inconspicuous constriction in the neck region (♂ HWI 38.5-*44.1*-50.7; ♀ HWI 29.5-*45.6*-63.4).

The web is moderately shallow (WDI ♂ 18.5-*22.2*-28.0; ♀ 20.4-*24.7*-29.5), and the web formula is very variable. In general, sectors C and D are the deepest, but occasionally A or E is deepest. The web extends out along the ventral side of the arms for about two-thirds of their length.

The arms are of medium length (ALI ♂ 56.2-*60.7*-65.9; ♀ 57.4-*62.0*-68.2) with the formula 3.2.4.1 or 4.3.2.1. The suckers are not large (SIn ♂ 8.6-*9.9*-11.1; ♀ 7.4-*9.3*-11.3) and are evenly spaced, separate, and erect. While the sucker index of the males is slightly larger than of the females, there are no specially enlarged suckers describable in the males.

In the males the third right arm is hectocotylized. It is a little shorter than its fellows (HcAI 72.0-*77.5*-83.5) and is bordered ventrally by a stout, inrolled membrane forming the spermatophoral groove. The ligula is small (LLI 5.0-*7.3*-10.0), narrow, and pointed, with inrolled margins and a distinct longitudinal groove. There are traces of transverse rugae. The calamus is short (CLI 15.0-*19.4*-26.3) and lies closely adhering to the ligula.

The gills are large. The outer demibranch bears about 12 to 16 lamellae, with a distinct average of 13-14.

The mandibles are as figured.

→

FIGURE 1. *Octopus selene,* new species: a, lateral view of holotype, 39.2 mm mantle length; b, upper and lower beaks; c, distal portion of hectocotylized arm of holotype; d, funnel organ; e, male reproductive tract; f, digestive tract; g, dorsal view of liver; h, radula; i, female reproductive tract.

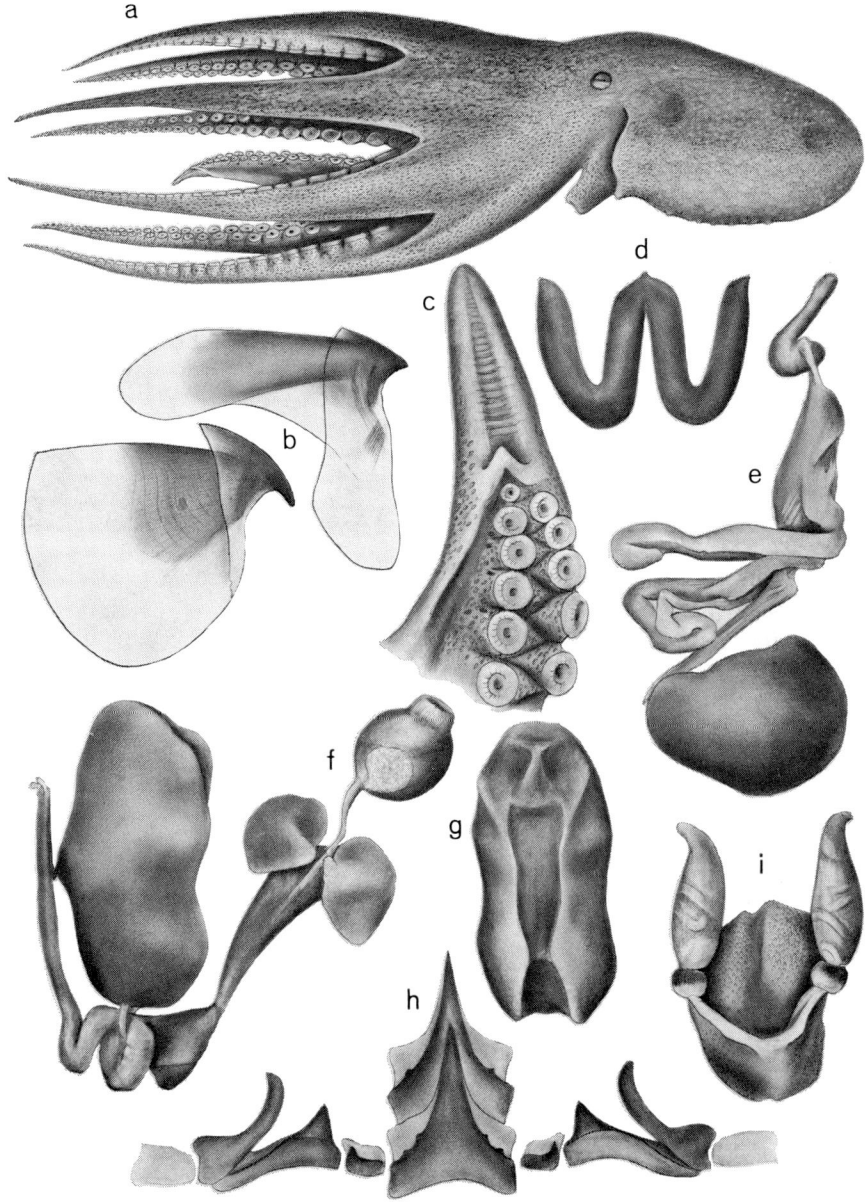

The radula is not distinctive. The rachidian teeth show an A^{3-4} seriation, although the cusps are small. The admedians are small, with a small, erect ectocone near the outer end of the tooth. The second lateral has a long base, with a broad, erect cusp near the inner end. The third laterals are of the long, slender, sabre type. The marginals are undeveloped and rectangular in shape.

The digestive tract was removed from the largest female. There are two large, leaflike posterior salivary glands and a large and conspicuous crop. The stomach is large and separated from the spiral caecum. The intestine does not appear differentiated and terminates with a pair of anal flaps. The ink sac is large and deeply set into the liver.

The female genitalia consist of a large ovary with numerous small eggs which appear nearly mature. They are 1.6 × 0.3 mm, with a short stalk. The proximal oviducts have a common pore and branch immediately, but are small in diameter. The oviducal gland is also small. The distal oviducts are very large and swollen, and both are filled with spermatophores.

The male genitalia are of the usual type. The penis is long and slender, with a short diverticulum. However, in some specimens the penial apparatus appears more as a loop, as shown in the figure.

The sculpture consists of low papillae or rugosities over the dorsal area of the mantle, head and arms. This is a variable feature, however, and some specimens are smooth, or nearly so.

The color in alcohol is light to dark reddish purple on the dorsum, light to pale brown on the ventrum, with several large dark spots or splotches on the mantle. Two are placed far posteriorly on the mantle and two more, one on each side, near the origin of the mantle aperture.

Type.—U. S. National Museum 577617.

Type-Locality.—PILLSBURY Sta. 501, 7°50.2′N, 79°50.5′W, northeast of Punta Mala, Panamá, in 68 meters with 10-ft OT, May 2, 1967.

Discussion.—This species does not appear to be closely related to any of the known eastern Pacific octopods. Although some of the octopods reported from the Pacific coast by early workers are so poorly known as to be almost unrecognizable, the present material appears to represent a new species, and it is here so considered.

The specimens were so numerous at two stations that it was calculated there were about one and a half octopuses per square meter. They were found on sandy and rocky bottom, along with such other animals as the mollusks *Calliostoma*, *Polinices*, and *Tellina*, the worm *Aphrodite* and amphinomids, the crustaceans *Squilla*, *Solenocera*, *Clibanarius*, and *Dardanus*, the echinoderm *Clypeaster*, and fishes of the genera *Scorpaena*, *Porichthys*, *Prionotus*, and *Symphurus*.

Remarks.—The specific name *selene* is from the Greek word *selene*, the moon, and is given in recognition of the first moon landing by the U. S. Astronauts Armstrong and Aldrin on July 20, 1969. This description was written on the day they lifted off from the moon's surface to return to the earth.

TABLE 5

MEASUREMENTS (IN MM) AND COUNTS OF 11 MALES AND 20 FEMALES OF *Octopus selene*, NEW SPECIES

	♂	♂	♂	♂	♂	♂	♂	♂	♂	♂
Mantle length	40.0	40.0	40.2	40.5	40.5	40.9	40.9	42.1	42.9	42.9
Mantle width	32.8	29.9	29.3	28.9	28.6	29.3	25.0	36.4	38.0	38.0
Head width	19.0	18.1	21.0	20.0	18.8	21.0	19.2	18.1	19.1	19.1
Arm length I	80.5	—	83.2	—	69.1	72.8	61.0	84.6	75.8	75.8
II	88.5	75.0	79.9	—	79.1	78.3	68.9	—	81.1	81.8
III	90.8	81.1	89.5	—	—	—	64.5	—	88.0	88.0
IV	86.0	88.9	82.4	—	86.5	89.0	63.8	86.9	83.1	83.1
Sucker diameter	4.0	4.2	3.9	3.5	4.6	4.2	3.5	4.0	3.2	3.2
Arm width	6.0	5.9	6.1	6.5	5.1	6.0	5.4	6.0	7.5	7.5
Gills	14	13	14	14	14	14	14	13	14	14
Web depth A	—	—	18.9	13.5	18.0	12.2	16.3	—	—	—
B	17.0	23.9	23.8	13.7	18.5	14.0	17.0	21.1	24.0	24.0
C	16.1	21.8	22.9	16.0	—	—	17.5	20.6	22.3	22.3
D	22.0	—	—	—	—	—	20.0	20.5	26.0	26.0
E	14.8	14.9	20.5	—	—	19.0	15.0	19.5	24.3	24.2

	♀	♀	♀	♀	♀	♀	♀	♀	♀	♀
Mantle length	43.0	43.2	43.7	44.0	44.1	45.4	45.7	47.9	55.2	58.1
Mantle width	30.9	29.2	31.9	26.8	31.5	26.7	31.1	37.4	43.2	46.0
Head width	17.9	19.9	18.9	18.0	21.0	19.2	29.0	19.1	26.9	25.8
Arm length I	70.1	91.0	75.9	78.1	86.2	76.9	85.8	84.2	124.2	126.0
II	78.1	103.1	—	89.5	93.4	88.0	89.7	93.9	137.9	136.9
III	—	108.3	86.3	89.1	91.4	90.1	88.1	95.0	—	142.3
IV	—	105.0	86.9	82.9	89.4	84.1	85.3	99.8	146.5	143.0
Sucker diameter	4.3	4.2	4.1	4.0	4.3	3.5	4.2	4.7	5.2	5.2
Arm width	5.8	7.0	6.0	7.1	6.1	5.2	6.0	7.1	7.4	11.3
Gills	14	13	13	12	15	14	13	14	13	14
Web depth A	17.4	19.2	19.2	17.4	18.3	23.5	16.0	20.0	24.8	25.0
B	15.9	—	—	19.5	20.0	17.1	18.3	21.9	26.4	20.4
C	—	23.2	17.5	23.2	25.3	21.8	16.1	24.1	32.5	31.9
D	—	24.8	21.9	15.8	17.0	—	18.2	24.5	31.9	30.8
E	—	19.0	12.0	17.8	18.1	16.8	19.1	16.5	27.1	26.8

TABLE 5 (Continued)
MEASUREMENTS (IN MM) AND COUNTS OF 11 MALES AND 20 FEMALES OF *Octopus selene*, NEW SPECIES

	♂	♂	♂	♂	♂	♂	♂	♂	♂	♂	♂
Mantle length	40.0	40.1	40.3	40.4	40.4	40.5	41.0	41.0	41.2	41.9	42.8
Mantle width	21.9	23.2	27.0	24.8	28.0	29.5	21.9	23.1	26.8	28.2	25.9
Head width	16.5	17.5	19.1	18.8	19.0	19.0	15.8	16.7	20.9	17.2	18.1
Arm length I	74.6	71.2	75.0	82.5	79.0	68.4	70.0	71.4	70.0	69.5	70.1
II	73.5	76.7	77.9	84.1	86.1	75.3	69.0	70.9	81.3	72.9	78.0
III	80.5	80.8	84.1	—	86.2	76.1	76.1	67.1	79.9	66.2	77.2
IV	78.1	76.1	76.8	83.2	84.1	73.4	73.0	—	72.9	66.5	77.0
Hect. arm length	—	67.5	64.1	68.0	62.0	56.8	58.6	56.6	63.1	55.4	61.2
Ligula length	—	4.2	5.1	6.0	6.2	3.8	4.0	3.3	4.7	4.0	5.6
Calamus length	—	1.1	0.8	1.0	0.6	1.0	0.77	0.6	0.9	0.6	1.28
Sucker diameter	3.7	3.5	4.5	3.5	4.5	4.1	3.6	4.1	4.2	4.0	4.5
Arm width	5.5	4.7	6.2	5.2	6.0	6.0	4.5	6.0	6.5	7.1	6.0
Gills	14	14	13	14	12	13	14	13	13	13	14
Web depth A	—	15.0	13.9	14.0	16.0	16.1	14.2	—	17.1	—	18.2
B	20.5	12.1	19.2	16.0	13.9	—	13.1	—	13.0	20.1	—
C	18.8	16.1	12.9	14.0	—	16.1	—	—	14.5	17.1	—
D	22.0	14.1	15.0	16.8	—	17.8	—	20.0	17.1	15.8	—
E	—	15.5	17.1	13.9	—	16.4	—	—	15.2	14.6	15.2

TABLE 6
INDICES OF BODILY PROPORTIONS OF 11 MALES AND 20 FEMALES OF *Octopus selene*, NEW SPECIES

Character	♂ Range & Mean	N	♀ Range & Mean	N
Mantle length	40.4–*40.9*–42.8	11	40.0–*44.6*–58.1	20
Mantle width index	49.7–*61.9*–72.8	11	58.8–*73.7*–86.4	20
Head width index	38.5–*44.1*–50.7	11	29.5–*45.6*–63.4	20
Arm length index	56.2–*60.7*–65.9	11	57.4–*62.0*–68.2	19
Mantle arm index	47.9–*51.8*–57.4	11	37.6–*47.4*–55.0	19
Arm width index	10.9–*14.1*–16.9	11	11.4–*14.8*–19.4	20
Web depth index	18.5–*22.2*–28.0	11	20.4–*24.7*–29.5	19
Sucker index	8.6– *9.9*–11.1	11	7.4– *9.3*–11.3	20
Hect. arm index	72.0–*77.5*–83.5	10	—	—
Ligula length index	5.0– *7.3*–10.0	10	—	—
Calamus length index	15.0–*19.4*–26.3	10	—	—

Octopus balboai, new species
Fig. 2, a-e

Material Examined.—HOLOTYPE: 1 ♂, mantle length 33.5 mm, from Chamé Point, Panamá, collected by Robert Tweedle, U. S. National Museum. USNM 577618.

Description.—The mantle is elongate-oval in outline, bluntly pointed posteriorly (MWI 68.7), and widest at about the midpoint. The head is only slightly narrower than the mantle width and is set apart from the mantle by a constriction in the neck region (HWI 65.6).

The mantle aperture is moderately narrow. The funnel is small but stout, and is free for about half of its length. The funnel organ is W-shaped, with thick apices and narrow, pointed limbs.

The web is shallow (WDI 23.9), with a formula of $D > E = C > B = A$, but there is little real difference between the sectors, which are about subequal.

The arms are long (ALI 68.7), stout at their bases (AWI 20.9), and taper to long, slender points. The web does not extend out the arms. The suckers are biserial, large (SIn 12.2), and crowded; none of them are conspicuously enlarged. In the male, the suckers of the distal third of the arms are curiously modified. The fleshy outer edge of each sucker bears a circlet of short, pointed papillae. These are visible only when the tips of the arms are immersed in liquid. The papillae then stand out plainly, giving the suckers a star- or flower-like appearance. This modification may be a sexual character similar to the modification of the arm tips in *Eledone*; this cannot be demonstrated until females have been collected and described.

The hectocotylized arm is short (HcAI 76.0), with a strongly inrolled ventral membrane forming the sphermatophoral groove. The ligula is small (LLI 7.5), slender, and pointed, with a distinct median groove. The calamus (CLI 40.0) is not long, but it is stout and stands conspicuously erect.

There are six lamellae on each outer demibranch of the gills.

The viscera were not dissected in the unique specimen. However, the penis is large and crescent-shaped, with the duct entering at about the midpoint. A few partially degenerated spermatophores were found in Needham's sac. One of these is illustrated. These spermatophores are unusual, somewhat resembling the armored spermatophores of *O. dollfusi* and certain *Eledone (Acantheledone)*. However, the "teeth" on the midportion of the spermatophore are more scalelike; the cap is also much simpler.

There is an ink sac deeply buried in the liver, its duct entering the rectum near the anal opening.

The color in alcohol is purplish brown, deeper on the dorsum of the mantle, with a few faint purplish flecks. The dorsum of the mantle, head, and arms is covered with small to large simple papillae, larger in the mantle. The eyes are surrounded by a group of closely set papillae, but there are no ocular cirrhi present.

Type.—U. S. National Museum 577618.

Type-Locality—Chamé Point, Panamá.

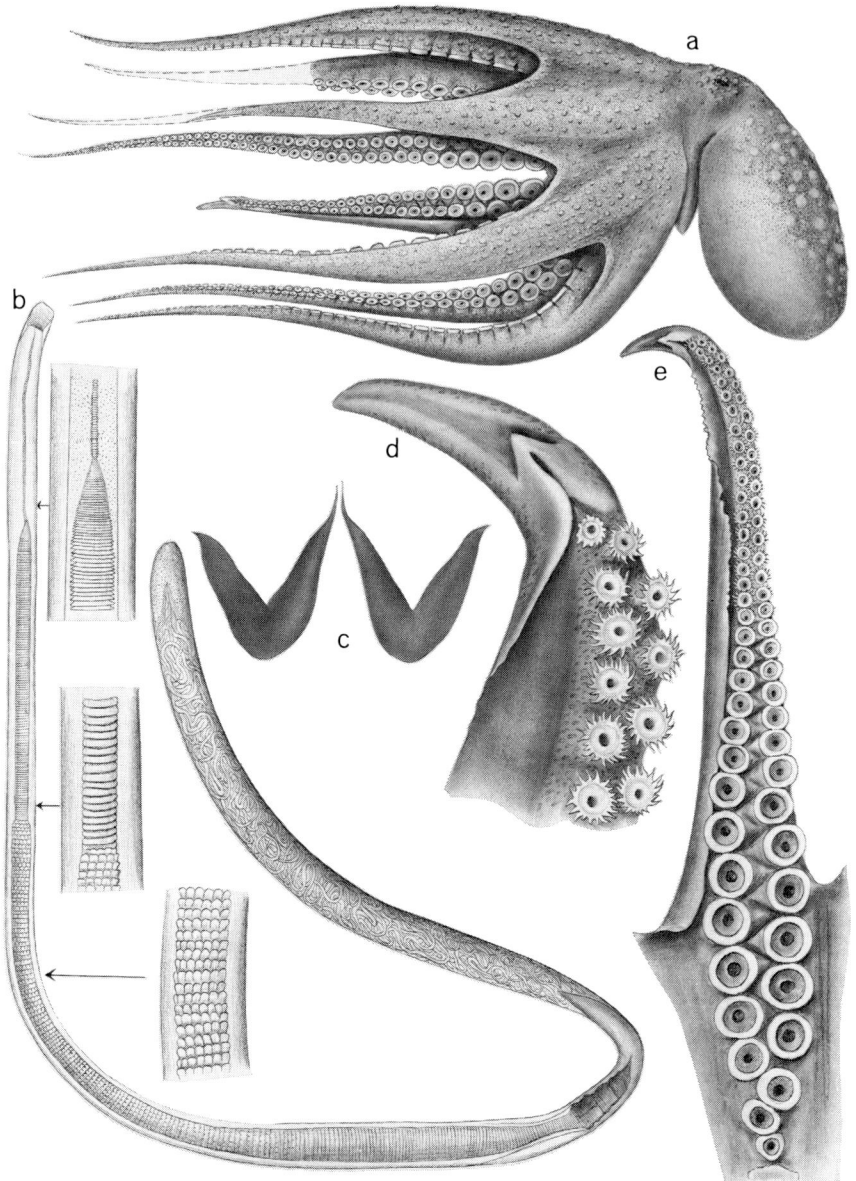

FIGURE 2. *Octopus balboai*, new species. Holotype, mantle length 33.5 mm: a, lateral view; b, spermatophore and enlarged sections; c, funnel organ; d, distal portion of hectocotylized arm; e, complete hectocotylized arm.

TABLE 7
MEASUREMENTS (IN MM) AND COUNTS OF THE HOLOTYPE OF *Octopus balboai*, NEW SPECIES

Sex	♂		Sucker diameter	4.1
Mantle length	33.5		Arm width	7.0
Mantle width	23.0		Gills	6
Head width	22.0		Web depth A	18.0
Arm lengths:	L.	R.	B	18.0
			C	20.0
I	75.0	45+	D	21.0
II	65.0	87.0	E	20.0
III	81.0	67.0		
IV	88.0	87.0		
Hect. arm length	67.0			
Ligula length	5.0			
Calamus length	2.0			

TABLE 8
INDICES AND FORMULAE OF BODILY PROPORTIONS OF THE HOLOTYPE OF *Octopus balboai*, NEW SPECIES

Mantle width index	68.7	Sucker index	12.2
Head width index	65.6	Hect. arm index	76.0
Arm length index	68.7	Ligula length index	7.5
Mantle arm index	38.1	Calamus length index	40.0
Arm width index	20.9	Arm formula	4.2.3.1
Web depth index	23.9	Web formula	D>E = C>B = A

Discussion.—This species is known only from the single specimen in the collections of the U. S. National Museum. In general appearance, it resembles a deep-water octopod. It is unfortunate that no other material is available.

Remarks.—The specific name is derived from the name of the first European to see the Gulf of Panamá—Vasco Nuñez de Balboa.

Octopus stictochrus, new species
Figs. 3, a-e; 4, a-e

Material Examined.—HOLOTYPE: ♀, mantle length 19 mm, PILLSBURY Sta. 535, 8°38.6′N, 78°51.9′W, between the Perlas Islands and Río Hondo, Panamá, in 31 meters with 10-ft OT, May 6, 1967. USNM 577619.

PARATYPE: ♀, mantle length 20 mm, PILLSBURY Sta. 492, 7°50.7′N,

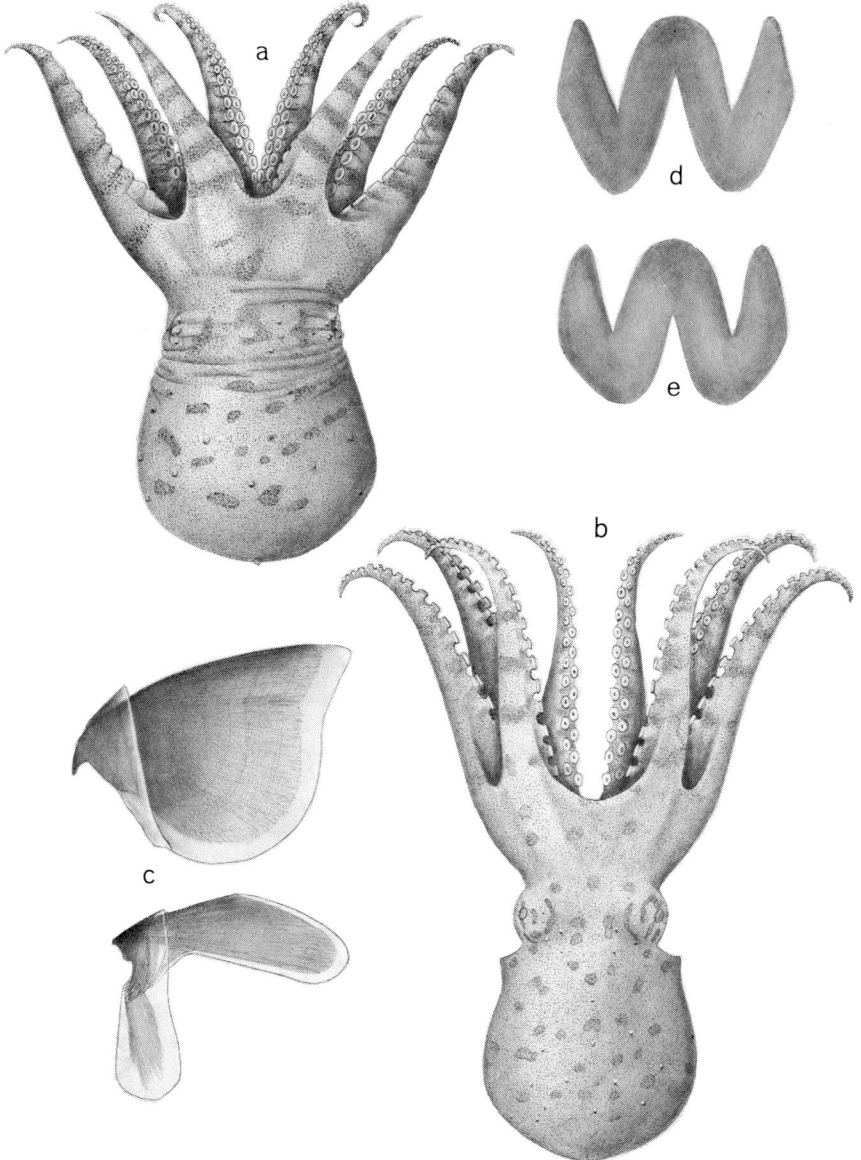

FIGURE 3. *Octopus stictochrus,* new species: a, dorsal view of holotype, mantle length 19 mm; b, dorsal view of paratype, mantle length 20 mm; c, upper and lower beaks of holotype; d, funnel organ of holotype; e, funnel organ of paratype.

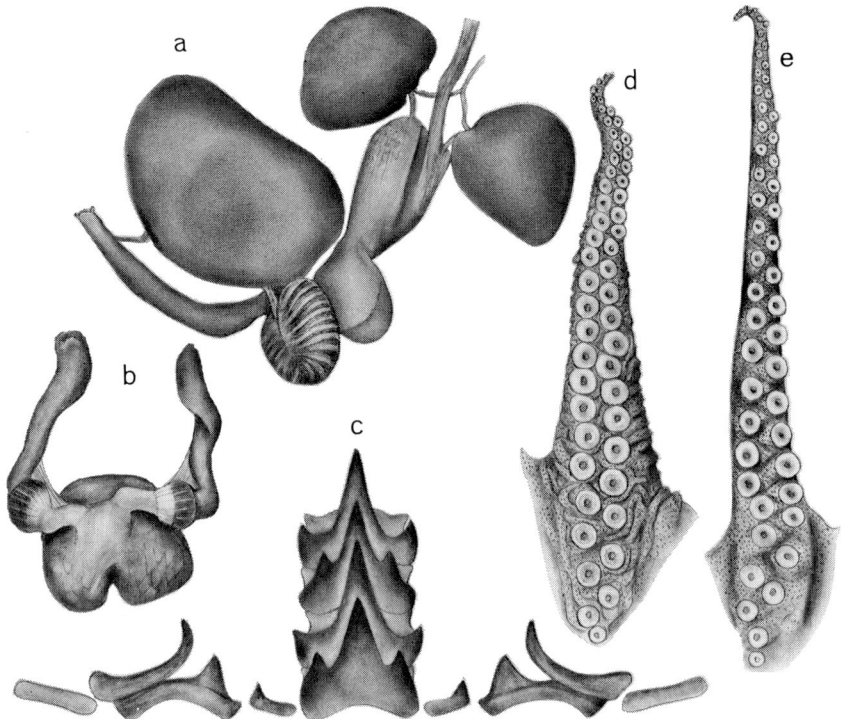

FIGURE 4. *Octopus stictochrus*, new species: a, digestive tract of paratype; b, female reproductive tract of paratype; c, radula of paratype; d, arm of holotype; e, arm of paratype.

80°09.8′W, off Isla Villa, Panamá, in 16-18 meters with 10-ft OT, May 2, 1967.

Description.—This is a small species of octopus, unfortunately represented only by female specimens, one of which seems to be gravid. The mantle is short, stout, and somewhat squarish posteriorly, decidedly wider near the end (MWI 90.0-100). There is no constriction between the head and the mantle, and the head is narrower than the mantle width (HWI 62.5-73.7).

The funnel is short and stout, with broad basal shoulders. It is free for about half its length. The funnel organ is W-shaped, with rather thick limbs.

The web is moderately deep (WDI 30.2-37.5) and subequal, with sector A slightly shallower than the others. The web formula is D.C.E.B.A. in both specimens. The web does not extend up the sides of the arms in either example.

TABLE 9
MEASUREMENTS OF THE HOLOTYPE AND PARATYPE OF *Octopus stictochrus*, NEW SPECIES

	Holotype ♀	Paratype ♀		Holotype ♀	Paratype ♀
Mantle length	19.0	20.0	Arm width	4.0	3.5
Mantle width	19.0	18.0	Gills	7	6
Head width	14.0	12.5	Total length	60.0	66.0
Arm length I	31.0	39.0	Web depth A	11.0	10.5
II	33.0	40.0	B	11.5	10.5
III	36.0	41.0	C	12.5	12.0
IV	36.0	41.0	D	13.5	12.5
			E	12.0	11.0
Sucker diameter	1.6	1.3			

The arms are short (ALI 60-62; MAI 48.7-52.8) and rather stout at the web margin (AWI 17.5-21.0). They show the formula 3 = 4.2.1, but are subequal with no disparity between pairs. The suckers are in two rows, small (SIn 6.5-8.4), and evenly spaced.

The gills are large and broad, but have only six or seven lamellae on each demibranch.

The mandibles are as figured and show no distinctive features.

The radula is somewhat unusual. The rachidian teeth show an A^{2-3} seriation. The admedians are small and low, but have a single, distinct, erect ectocone near the outer end of the tooth. The second lateral is long with a subterminal endocone. The third lateral is sabrelike. The marginals are well developed, long, and slender.

The digestive tract is of the usual type. The second salivary glands are large and leaflike. The crop is well developed. The spiral caecum is larger than the stomach, dark gray, and marked with white transverse lines. The large intestine is undivided. The ink sac is set into the liver.

The female genitalia were dissected. The ovary is small and undeveloped. The proximal oviduct is short and stout. The oviducal gland is dark gray, with white ribbing; the distal oviducts are large.

The color in alcohol is light mauve brown dorsally and ventrally on the mantle, head, and arms. On the mantle there are about six or seven transverse, irregular rows of large dark spots. These extend midway down the sides, but are absent on the ventral surface. These are somewhat fused around the eyes. The spots extend out onto the arm bases, but, just proximal of the border of the web, there appear bands that, beyond the web, extend across the arms from sucker base to sucker base. There are

TABLE 10
INDICES OF BODILY PROPORTIONS OF HOLOTYPE AND PARATYPE OF *Octopus stictochrus*, NEW SPECIES

	Holotype	Paratype
Mantle width index	100.0	90.0
Head width index	73.7	62.5
Arm length index	60.0	62.0
Mantle arm index	52.8	48.7
Arm width index	21.0	17.5
Web depth index	37.5	30.2
Sucker index	8.4	6.5

about six to nine bands across all arms in the specimen from Sta. P-535 (Fig. 1,a); the other specimen (Fig. 1,b) is not distinctly banded, the bands being replaced by more irregular brown spots. There is no evidence of any sculpture; the surface is smooth.

Type.—U. S. National Museum 577619.

Type-Locality.—PILLSBURY Sta. 535, 8°38.6'N, 78°51.9'W, between the Perlas Islands and Río Hondo, Panamá, in 31 meters with 10-ft OT, May 6, 1967.

Discussion.—This species seems to be distinctive, and I know of no other species which resembles it, either along the west coast of the Americas or in the Pacific islands. The characteristic color makes it easily recognizable.

These two specimens were found in the same tows with *O. chierchiae*. A brief listing of other benthic animals shared in both tows is given in the notes regarding the latter species.

The name *stictochrus* is derived from the Greek *stictos* and *chróos* meaning "with spotted skin."

Octopus chierchiae Jatta, 1889

Octopus chierchiae Jatta, 1889: 64; 1899: 19, pl. 1, figs. 3-14.—Robson, 1929: 152.—Voss, 1968: 652, fig. 1, b, fig. 2, g-h, fig. 3, f-j, fig. 4, e-f.

Material Examined.—1 ♀, mantle length 29 mm, PILLSBURY Sta. 492, 7°50.7'N, 80°09.8'W, off Isla Villa, Panamá, in 16-18 meters with 10-ft OT, May 2, 1967.—1 ♀, mantle length 18.5 mm, PILLSBURY Sta. 535, 8°38.6'N, 78°51.9'W, between the Perlas Islands and Río Hondo, Panamá, in 31 meters with 10-ft OT, May 6, 1967.

This beautiful small species of banded octopus was taken at only two stations, both times in company with a small spotted octopus, *Octopus stictochrus*. It was described and figured by Jatta from material collected by the VETTOR PISANI from the Gulf of Panamá. It was figured by Voss (1968) and compared in detail with its twin Caribbean species, *Octopus zonatus*. Both of the above specimens were immature.

Field notes taken at the time (Voss, 1967) show that the two localities, although at opposite ends of the gulf, shared a number of genera: *Conus, Distorsio, Chione, Tellina,* and *Turris* among mollusks; *Hepatus, Raninoides, Sicyonia,* and *Squilla* among crustaceans; and *Symphurus, Porichthys,* and genera of bothids and gobies among the fishes.

This species is now known to occur from the Gulf of Panamá to El Salvador.

? *Octopus oculifer* (Hoyle, 1904)

?*Octopus bimaculatus* Verrill, 1883: 122 (specimen listed from Panamá and San Salvador; not *O. bimaculatus* Verrill, 1883, from California).
?*Polypus oculifer* Hoyle, 1904: 14, pl. 4, figs. 3, 4.

Material Examined.—1 ♀, mantle length 10.3 mm, Isla Saboga, Canal de Santiago, February 10, 1948, Coll. Paul S. Goltsoff, U. S. National Museum.

A small specimen of ocellated octopus was found in the U. S. National Museum collections of Panamanian cephalopods. The specimen is an immature female. The skin is rough, with numerous small tubercles and scattered, tall, erect papillae; a cirrus occurs over each eye. The ocellus is composed of a light outer ring, a dark band inside, and a pale center. There are about ten lamellae on the outer demibranch of the gill.

An additional specimen of what appears to be the same species was taken by Charles E. Dawson on the Pacific coast of Costa Rica at Guanacaste, Playa el Coco, with fish poison from a tide pool. This is a young male wtih a mantle length of 14.0 mm.

The identity of these specimens must remain in doubt until all available specimens are studied and the group revised. Pickford (1945) and Pickford & McConnaughey (1949) have pointed out the existence of an unnamed species of ocellated octopus in the Panamanian region. Examination of the two specimens listed above shows that this species is very close to Hoyle's *Polypus oculifer* from the Galapagos Islands, and, until such time as the necessary studies have been made, I am referring the present specimens to this poorly known species.

? *Octopus vulgaris* Cuvier, 1797

Material Examined.—2 ♀, mantle length 18.0-31.0 mm, Bahia Santelmo, Isla del Rey, Perlas Islands, Panamá, with fish poison along rocky shore, August 1, 1968. Coll. Charles E. Dawson.

The two very granulose specimens are tentatively referred to this species.

Both are young females, and no good characters are available. The gills have about ten lamellae per outer demibranch, and the funnel organ is W-shaped. The smaller specimen has a considerably elongated mantle; the larger specimen is more usual in appearance.

Octopus sp.

Material Examined.—1 ♀, mantle length 32.0 mm, PILLSBURY Sta. 531, 8°25.5'N, 79°10.7'W, west of the Perlas Islands, Panamá, in 57-64 meters with 10-ft OT, May 6, 1967.

The single immature female has no distinctive features sufficient to yield an identification and does not seem referable to any of the known species of the area.

Octopus Eggs

Material Examined.—Clutch of about 100 eggs in venerid shell from PILLSBURY Sta. 535.

A clutch of about 100 eggs measuring 7.0 × 3.0 mm was recovered in a venerid valve at this station. The eggs are in a single layer attached terminally by a short thin stalk. At the present state of our knowledge of the Panamic octopods, it is not possible to assign these with certainty to any of the known Panamic species. Robson (1929:199) remarked that the eggs of *O. digueti* Perrier & Rochebrune (1894) from the Gulf of California are laid in bivalve shells in groups of two to five with the thin stalk attached to the shell in the middle of a circular or oval dark patch. The same is true of the present eggs, the dark patches being particularly noticeable against the white surface of the shell.

Euaxoctopus panamensis, new genus and new species
Figs. 5, a-c; 6, a-j

Material Examined.—HOLOTYPE: ♂, mantle length 26.0 mm, PILLSBURY Sta. 493, 7°39.5'N, 80°00.7'W, north of Isla Iguana near Punta Mala, Panamá, in 37-33 meters with 10-ft OT, May 2, 1967. USNM 577620.

PARATYPES: 2 ♂, mantle length 21.0-32.0 mm, 2 ♀, mantle length 20.0-29.0 mm, PILLSBURY Sta. 493, with holotype.—1 ♂, mantle length 25.0 mm, PILLSBURY Sta. 533, 8°45.2'N, 79°10.3'W, northwest of the Perlas Islands, Panamá, in 37-33 meters with 10-ft OT, May 6, 1967.—1 ♀, mantle length 14.5 mm, seaward of Teacher Ferry Bridge, canal bottom, from dredge flume, Panama Canal, March 16, 1967. USNM 576454.

Description.—This unusual species of octopod is represented in the collections by six specimens, most of which, with the exception of broken arms, are in excellent condition.

The mantle is ovoid, somewhat slender and bluntly pointed posteriorly

(MWI 47.5-*53.6*-59.3). There is a strong constriction between the mantle and the head. The neck region is long and conspicuous; the head is narrow with small, protuberant eyes (HWI 31.7-*36.8*-42.5).

The funnel is long and slender and free for about ⅓ to ½ of its length. The funnel organ is VV-shaped, with long, slender arms.

The web is very shallow, but, because of the frequent mutilation of the arms, the depth is difficult to measure. Only two specimens (Nos. 1 and 5) had intact long arms. These yielded web depth indices of 8.7 and 8.8, respectively. The web sectors were about equal in depth, with E usually being the shallowest; in one specimen, however, E was the deepest.

The arms are long (for three nearly complete specimens, ALI 80.0-*81.6*-84.0; MAI 15.8-*18.3*-20.8) and slender (AWI 11.9-*13.9*-15.4). Since the AWI is calculated as a percentage of the mantle length, it does not in this case portray the true nature of the slenderness of the arms. Arms II and III seem regularly to be longest, and IV the shortest, but because of arm breakage and partial regeneration the arm formula is not certain. The suckers are in two rows, small (SIn 5.2-*6.3*-8.0), widely spaced, and erect. There was no indication of specially enlarged suckers in the males.

In the three males the third left arm is hectocotylized. This arm is much shorter than its fellows (HcAI 25.6-*26.8*-28.0) and bears a small, slender ligula at its tip (LLI 6.0-*7.6*-9.2). The ligula is narrow and pointed, with a broad groove, strongly inrolled margins, and numerous fine transverse lines. The calamus is short, stout, and about ⅓ to ½ the length of the ligula (CLI 32.4-*46.1*-54.0). The arm is bordered ventrally by an expanded web, but there is no distinct spermatophoral groove.

The gills contain about 11 lamellae per demibranch, but two specimens have 12 and 13, respectively (gills 11-13).

The mandibles are as figured.

The radula is not unusual nor distinctive. The rachidian teeth have an A^2 seriation, with a single large cusp on each side. The admedian is short and stout, with a blunt ectocone. The second laterals are long, with a heavy broad endocone. The third laterals are sabrelike. The marginals are rectangular and poorly developed.

The digestive tract was removed from the largest male (No. 4). It shows a number of unusual features. The posterior salivary glands are of unequal size; the left gland is about twice the size of the right, and both

→

FIGURE 5. *Euaxoctopus panamensis*, new genus and species: a, lateral view of holotype, mantle length 26 mm; b, hectocotylized arm of holotype; c, hectocotylus of holotype; d, dorsal view of mantle and head, showing size and shape of ocellus.

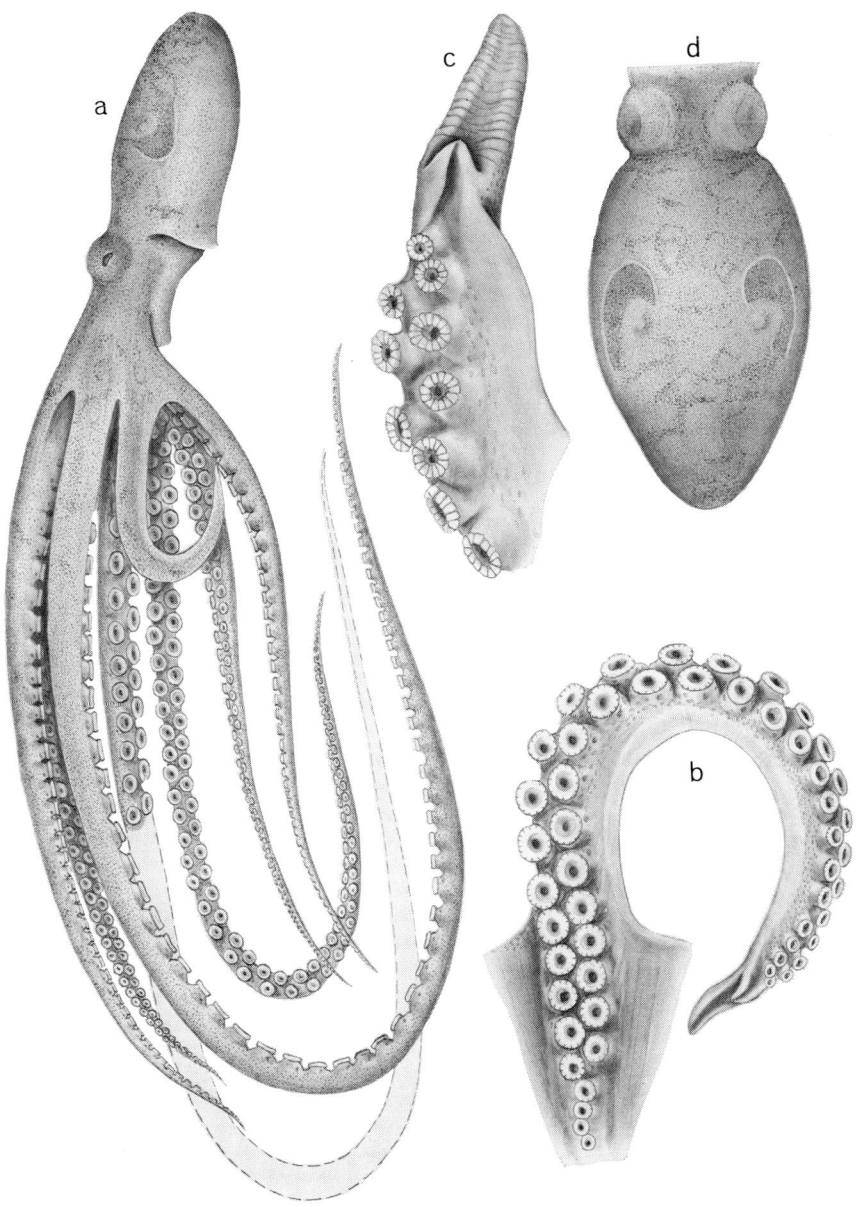

appeared to be fused together to form a cap dorsally and anteriorly over the crop. The glands could be teased apart, but were actually fused though they had separate ducts.

The crop is large and well developed, with the oesophagus entering dorsally. From the crop, the oesophagus runs posteriorly along the dorsal surface of the liver to become a large, soft stomach fused to a small, striated, spiral caecum. The stomach–spiral caecum complex possesses two poorly developed ducts leading to the posterior end of the liver. The intestine is partially convoluted; the terminal pore is protected by lateral winged flaps. The liver is very elongate and rather slender and posteriorly is broadly cleft with large out-turned flaps. There is a well-developed ink sac partially buried in the liver, with a duct entering the rectum just proximal of the anal opening.

The female genitalia were dissected in specimen No. 3. The ovary was large and contained a mass of nearly ripe, small eggs about 1.4×0.5 mm, with small stalks. The proximal oviducts have a common origin and then divide to lead to large, grayish black oviducal glands. The distal oviducts are short.

The male genitalia were also dissected, but showed no unusual features. The penis is small, swollen, elongate oval, with the duct coming in near the anterior pore. In specimen No. 1 there were a number of fully formed spermatophores, one of which is illustrated. This shows a number of unusual features, such as the lack of a cement body and the odd terminal region.

The color in alcohol is a light grayish tan, with darker mottling on the dorsum of the head, arms, and mantle. The most prominent feature is the presence of a peculiar ocellus (?) on each side of the mantle, slightly dorsal and at about the midpoint between the neck region and the end of the body. The ocellus consists of a light bluish or pinkish crescentic line, the anterior end slightly more turned in, enclosing a darkly pigmented semicircular splotch. In life this ocellus was very conspicuous, but did not have the luminous glow of the usual ocellus. It disappeared at first on preservation in formalin, but reappeared after a short period and is conspicuous in alcohol.

The dorsum of the head, arms, and mantle is variously sculptured. The skin is nearly smooth in some, rugose in others, and has widely separated papillae in others. It can best be described as conspicuously sculptured but variable.

→

FIGURE 6. *Euaxoctopus panamensis*, new genus and species: a, digestive tract; b, dorsal view of liver; c, female reproductive tract; d, male reproductive tract; e, penis; f, upper and lower beaks; g, radula; h, funnel organ; i, funnel organ; j, spermatophore.

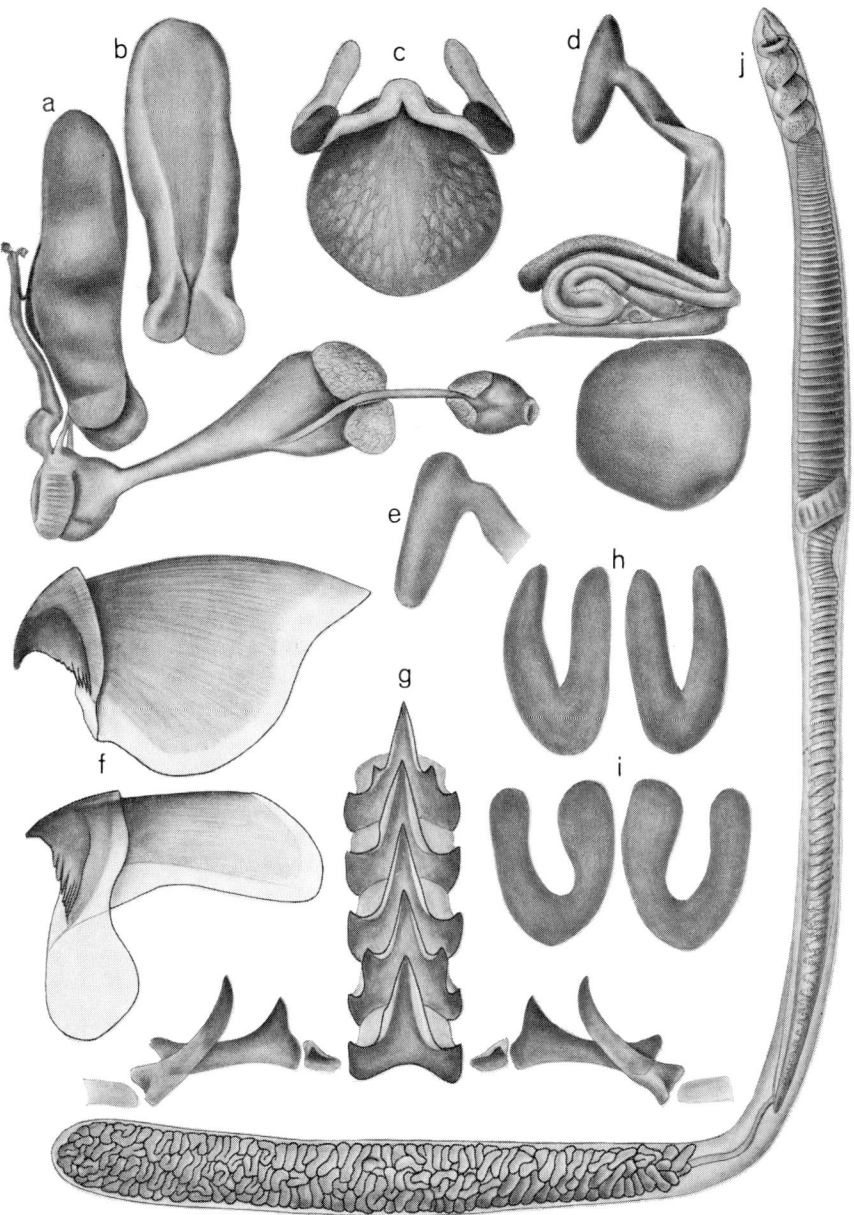

TABLE 11

MEASUREMENTS (IN MM) AND COUNTS OF THE HOLOTYPE AND FIVE PARATYPES OF *Euaxoctopus panamensis*, NEW SPECIES

	1		2	3		4		5		6	
Sex	♂		♀	♀		♂		♂		♂	
Mantle length	25.0		20.0	29.0		32.0		26.0		21.0	
Mantle width	14.0		11.0	14.5		19.0		14.0		10.0	
Head width	9.0		8.5	9.2		12.0		8.9		8.2	
Arm length I	89.0	87.0	67+	91.0	18+	107.0	97+	102.0	95.0	90.0	77.0	74.0
II	80.0R	84.0	37R	57R	25+	10+	154+	19+	164.0	65+	14+	15+
III	38.0	136.0	42R	21+	20+	29+	26+	70+	42.0	108.0	34.0	56+
IV	69.0	69.0	67.0	62.0	74.0	74.0	100.0	81+	70.0	65.0	54.0	54.0
Hect. arm length	38.0		—	—		—		42.0		34.0	
Ligula length	3.5		—	—		—		2.5		2.4	
Calamus length	1.1		—	—		—		1.3		1.3	
Sucker diameter	1.5		1.6	1.6		1.8		2.0		1.1	
Arm width	3.5		3.0	3.5		4.0		4.0		2.5	
Gills	11		11	12		—		12		13	
Web depth A	9.5		8.0	11.0		—		9.0		8.5	
B	11.5		8.0	11.0		—		11.0		8.5	
C	11.0		9.0	11.5		—		11.0		8.0	
D	11.0		9.0	11.0		—		10.0		8.0	
E	12.0		8.0	11.0		—		8.0		7.5	

R = regenerating arm tip.
+ = arm broken, with part missing.

TABLE 12
INDICES OF BODILY PROPORTIONS OF HOLOTYPE AND PARATYPES OF *Euaxoctopus panamensis*, NEW SPECIES

Character	N	Range & Mean
Mantle width index	6	47.5–*53.6*–59.3
Head width index	6	31.7–*36.8*–42.5
Arm length index	3	80.0–*81.6*–84.0
Mantle arm index	3	15.8–*18.3*–20.8
Arm width index	6	11.9–*13.9*–15.4
Web depth index	2	8.7–*8.75*– 8.8
Sucker index	6	5.2– *6.3*– 8.0
Hect. arm length index	2	25.6–*26.8*–28.0
Ligula length index	3	6.0– *7.6*– 9.2
Calamus length index	3	32.4–*46.1*–54.0

Type.—U. S. National Museum, 577620.

Type-Locality.—PILLSBURY Sta. 493, 7°39.5′N, 80°00.7′W, north of Isla Iguana near Punta Mala, Panamá, in 37-33 meters with 10-ft OT, May 2, 1967.

Discussion.—The appearance of a third genus of octopodines with the hectocotylus on the third left arm seems needful of some special comments. All of the known octopodines, with the exception of *Scaeurgus* and *Pteroctopus,* have the hectocotylus on the third right arm. Robson (1929) discussed this feature, both in his general considerations and under each of the two genera mentioned above. The genera *Scaeurgus* and *Pteroctopus* have little in common and do not seem related.

The present genus seems to have developed in shallow water as has *Scaeurgus*, but again, no close relationship seems evident. It is easily distinguished from *Scaeurgus* by the elongate body, VV-shaped funnel organ, the very long arms, shallow web, and the ink sac deeply involved in the liver.

As known, it represents a genus endemic to the Gulf of Panamá, with no representative or close ally in other regions of the world.

Remarks.—The name *Euaxoctopus* is derived from the Greek *euaxos*, "easily broken," and refers to the tendency for the arms to break off the brachial crown.

Family Argonautidae

Argonauta pacificus Dall, 1872

Argonauta pacificus, Keen, 1958: 514, fig. 3.

Material Examined.—1 ♀, alive with shell, eggs, and hatching larvae, mantle length 43.0 mm, PILLSBURY Sta. 529, 8°00.7′N, 79°11.8′W, south of Isla de San José, Perlas Islands, Panamá, in 84 meters with 10-ft OT, May 6, 1967.

The single female was taken along with eggs and hatching larvae. The shell was apparently crushed in the open trawl. The animal checks out very favorably with the brief description of this species given by Dall (1908). However, a full and detailed study of *pacificus* and *argo* is needed to determine whether they are conspecific, as some authors have considered them, or if they are indeed separate species.

Argonauta cornutus Conrad, 1854

Argonauta cornutus, Keen, 1958: 514, fig. 1.

Material Examined.—2 shells, 1 broken, 1 length 45.0 mm, PILLSBURY Sta. 533.—3 broken shells, PILLSBURY Sta. 500.—1 broken shell, PILLSBURY Sta. 519.—1 immature shell, length 14.0 mm, PILLSBURY Sta. 561. —1 shell, length 25.0 mm, PILLSBURY Sta. 530.—1 broken shell, PILLSBURY Sta. 549.—1 broken shell, PILLSBURY Sta. 555.—1 broken shell, PILLSBURY Sta. 532.—1 shell, length 42.0 mm, PILLSBURY Sta. 531.— 1 shell, length 42.0 mm, PILLSBURY Sta. 513.—1 broken shell, PILLSBURY Sta. 496.—3 shells, lengths 21.0, 26.0, and 32.0 mm, PILLSBURY Sta. 529. —1 broken shell, PILLSBURY Sta. 517.

Argonauta cornutus was very briefly and inadequately described by Conrad in 1854. The description was supplemented by a figure which shows several characters of importance, in particular the large, scattered tubercles on the midportion of the keel. The large series taken by the PILLSBURY was all from dead animals and was taken from the bottom in otter trawl catches, thus no soft parts are available. The shells, however, were carefully compared with those of whole animals from billfish stomachs from off Ecuador and Peru and were found to be similar in all respects.

A. cornutus seems best characterized by the few radial ribs, the presence of fine sharp tubercles or papillae over the sides of the shells, the few, rather sharp, large carinal tubercles on each edge, the *convex* carinal surface, and the few, large, blunt tubercles on the carinal surface between the two rows of carinal boundary tubercles.

Argonauta nouryi Lorois, 1852

Argonauta nouryi, Keen, 1958: 514, fig. 2.

Material Examined.—3 shells, partially broken, lengths 37.0, 46.0, and 57.0 mm, PILLSBURY Sta. 512.—1 shell, length 30.0 mm, PILLSBURY Sta. 373.—2 shells, 1 broken, 1 length 35.0 mm, PILLSBURY Sta. 519.—1 shell, length 42.0 mm, PILLSBURY Sta. 529.—1 broken shell, PILLSBURY Sta. 553.

Argonauta nouryi is a distinctive species fairly well represented in the PILLSBURY material. All of the shells were from dead animals taken on the bottom with the otter trawl. The shells are longer than in other species of *Argonauta*, the ribs are more numerous, there are no distinct tubercles marking the edges of the carinal area; the carina is wide, very convex, and covered by numerous, small, blunt tubercles formed by the crisscrossing of the ribs. Unfortunately, no soft parts are available.

Sumario
Cefalópodos Colectados por el Barco de Investigaciones John Elliott Pillsbury en el Golfo de Panamá en 1967

Se revisa la fauna de cefalópodos del Golfo de Panamá, basándose principalmente en colecciones hechas por el barco de investigaciones John Elliott Pillsbury durante 1967. Se dan reportes de 812 ejemplares, de los cuales cuatro especies y un género son nuevos para la ciencia: *Octopus selene, O. balboai, O. stictochrus* y *Euaxoctopus panamensis*. Ahora se conocen 26 especies en el Golfo, de las cuales 11 constituyen reportes registrados aquí por primera vez. Se discuten las relaciones de la fauna del Golfo de Panamá con las del Pacífico oriental y sudoeste del Mar Caribe.

LITERATURE CITED

Berry, S. Stillman
    1911. A note on the genus *Lolliguncula*. Proc. Acad. nat. Sci. Philad., 1911: 100-105, 1 pl., 7 text-figs.
    1929. *Loliolopsis chiroctes*, a new genus and species of squid from the Gulf of California. Trans. S Diego Soc. nat. Hist., 5(18): 263-282, pls. 32-33, figs. 1-9.

Conrad, T. A.
    1854. Monograph of the genus *Argonauta* Linné with descriptions of five new species. J. Acad. nat. Sci. Philad., 2(2): 331.

Dall, W. H.
    1908. The Mollusca and the Brachiopoda. Reports on the dredging operations off the west coast of Central America... XXXVI, and Reports on the scientific results of the expedition to the eastern tropical Pacific... XIV. Bull. Mus. comp. Zool. Harv., 43(6): 203-487, pls. 1-20.

Hoyle, W. E.
    1904. Reports on the Cephalopoda... U. S. Fish Comm. Steamer "Albatross" 1899-1900. Bull. Mus. comp. Zool. Harv., 43(1): 1-71, 10 pls.

Jatta, Giuseppe
    1889. Elenco dei Cefalopodi della "Vettor Pisani." Boll. Soc. Nat. Napoli, Ser. 1, 3(1): 63-67.
    1899. Sopra alcuni Cefalopodi della Vettor Pisani. Boll. Soc. Nat. Napoli, Ser. 1, 12(1): 17-32, figs. 1-29.

Joubin, L.
    1929. Notes préliminaires sur les céphalopodes des croisières du Dana (1921-1922). Octopodes—1re Partie. Annls Inst. océanogr., Monaco, New Series, 6(4): 363-394, 23 text-figs.

1931. Notes préliminaires sur les céphalopodes des croisières du Dana (1921-1922). 3ᵉ Partie. Annls Inst. océanogr., Monaco, New Series, *10*(7): 169-211, 48 text-figs.

KEEN, A. MYRA
1958. Sea shells of tropical west America. Marine mollusks from Lower California to Columbia. Stanford University Press, Stanford, California, 626 pp.

PERRIER, E. AND A. DE ROCHEBRUNE
1894. Sur un Octopus nouveau. C. r. Acad. Sci., Paris, *118:* 770.

PFEFFER, GEORG J.
1912. Die Cephalopoden der Plankton-Expedition. Ergebn. Atlant. Ozean Planktonexped. Humboldt-Stift., *2:* 1-815, atlas of 48 pls.

PICKFORD, GRACE E.
1945. Le Poulpe américain; a study of the littoral Octopoda of the western Atlantic. Trans. Conn. Acad. Arts Sci., *36*(July): 701-811, 19 pls.

PICKFORD, GRACE E. AND BAYARD H. MCCONNAUGHEY
1949. The *Octopus bimaculatus* problem; a study in sibling species. Bull. Bingham oceanogr. Coll., *12*(4): 1-66, 28 text-figs.

ROBSON, GUY COBURN
1929. A monograph of the Recent Cephalopoda. ... Part 1. Octopodinae. *1:* 1-236, 88 figs., 7 pls.
1932. A monograph of the Recent Cephalopoda. ... Part 2. The Octopoda, excluding the Octopodinae. *2:* 1-359, 79 figs., 6 pls.

ROPER, CLYDE F. E.
1968. Preliminary descriptions of two new species of the bathypelagic squid *Bathyteuthis* (Cephalopoda; Oegopsida). Proc. biol. Soc. Wash., *81:* 161-172, 7 pls.
1969. Systematics and zoogeography of the worldwide bathypelagic squid *Bathyteuthis* (Cephalopoda: Oegopsida). Bull. U.S. natn. Mus., No. 291: 1-210, 74 text-figs., 12 pls.

THORE, SVEN
1949. Investigations of the "Dana" Octopoda. Dana Rep., No. 33: 1-85, 69 figs.

VERRILL, A. E.
1883. Description of two species of octopus from California. Bull. Mus. comp. Zool. Harv., *11*(6): 117-124, 4 pls.

VOSS, GILBERT L.
1963. Cephalopods of the Philippine Islands. Bull. U. S. natn. Mus., No. 234: 1-180, 36 text-figs., 4 pls.
1967. Narrative of R/V JOHN ELLIOTT PILLSBURY Cruise P-6703 in the Gulf of Panamá, April 29–May 11, 1967. Institute of Marine Sciences, University of Miami, 30 pp., 1 map. (Processed report.)
1968. Biological Investigations of the Deep Sea. 39. Octopods from the R/V PILLSBURY southwestern Caribbean cruise, 1966, with a description of a new species, *Octopus zonatus*. Bull. Mar. Sci., *18*(3): 645-659, 4 text-figs.
In Press. Biological results of the University of Miami Deep-Sea Expeditions. A review of the cephalopods of the Gulf of Guinea.

BIOLOGICAL RESULTS OF THE UNIVERSITY OF MIAMI DEEP-SEA EXPEDITIONS. 77.
MOLLUSKS FROM THE GULF OF PANAMA COLLECTED BY R/V JOHN ELLIOTT PILLSBURY, 1967[1]

AXEL A. OLSSON
Coral Gables, Florida

ABSTRACT

Forty-one species of mollusks from the Panamic-Pacific faunal province are described. The following new species are described and figured: *Chama corallina, Tellina argis, Terebra argosyia, Glyphostoma bayeri, Carinodrillia dariena, Engina macleani, Phymorhynchus speciosus, Austrotrophon panamensis, Natica inexpectans, Calliostoma pillsburyae, C. insignis, C. joanneae, C. decipiens, Turcica panamensis,* and *Pectinodonta gilbertvossi.* One new genus is proposed, *Bayerius* (type-species, *Fusinus fragilissimus* Dall), and the new subgenus *Cantharus* (*Muricanthurus*) (type-species, *Pseudoneptunea panamica* Hertlein & Strong), both in the family Buccinidae. Relationships in the broad family Turridae are discussed, and the radular characters of several genera are described and illustrated. It is shown that *Strombinoturris crockeri* Hertlein & Strong belongs to the genus *Brachytoma* Swainson and is a junior synonym of *Brachytoma stromboides* (Sowerby). Dall's *Cocculina nassa* is shown to be a species of *Pectinodonta* allied to *P. gilbertvossi,* n. sp.

INTRODUCTION

Together with Dr. Frederick M. Bayer of the Rosenstiel School of Marine and Atmospheric Sciences of the University of Miami, the author has been engaged in a general study of the molluscan collections obtained during the cruise of the R/V JOHN ELLIOTT PILLSBURY in the Gulf of Panama in 1967 and of the yacht ARGOSY in 1961 from Panama southward along the coast of Ecuador to Isla la Plata. Hitherto, our principal information on Panamic-Pacific mollusks has been largely restricted to shore forms readily obtainable in tidal waters; therefore the PILLSBURY and ARGOSY mollusks from offshore stations have added much needed information concerning abundance and distribution, data which later may be useful in the assessment of changes in ecology which may result through the construction of a sea-level canal across the Isthmus of Panama. The original plan of the PILLSBURY cruise called for midwater and bottom trawling throughout the gulf, not only on the shelf and slope, but also beyond the Continental Shelf. Unfortunately, due to mechanical difficulties, only a single bottom haul was taken in deep water (station 526) south of the fault scarp which marks the

---

[1] Contribution No. 1310, from the University of Miami, Rosenstiel School of Marine and Atmospheric Sciences. This paper is one of a series resulting from the National Geographic Society–University of Miami Deep-Sea Biology Program.

south edge of the shelf. The results of this haul were so spectacular in all groups that further exploration in the general area would be desirable.

In the bottom operations of the PILLSBURY, the standard 10- and 40-foot otter trawls were used, so that mainly medium to large sized mollusks were taken, most of the smaller ones being lost. It is hoped that in any future operation in the gulf, especially near the Pearl Islands, some dredging will be made for the recovery of the smaller forms which are known to abound in that area.

It is the purpose of this paper to record some of the most interesting highlights from the collection of mollusks obtained by R/V PILLSBURY, with a few from that of the ARGOSY, and the unexpected discovery of some forms hitherto unknown in Panamic waters, along with the description of some new species, comments on others, and some observations on radulae. A list of the molluscan species obtained in the Bay of Panama during the PILLSBURY cruise will be given in another paper, along with other invertebrates, etc. A separate report on the ARGOSY mollusks will appear elsewhere.

ACKNOWLEDGMENTS

The opportunity to work with the molluscan collection secured by the PILLSBURY and ARGOSY cruises in Panamic and Ecuadorian waters has been a special privilege for which I am deeply indebted to Dr. Gilbert L. Voss, who was Chief Scientist aboard Research Vessel JOHN ELLIOTT PILLSBURY during its cruise in the Gulf of Panama and under whose direction the collections were so carefully preserved, and especially to Dr. Frederick M. Bayer, in charge of the invertebrate collections at the School, for our close cooperation in all phases of this study. Also, I am pleased to thank Mrs. Jo Anne Romfh, former assistant in the Deep-Sea Biology Project, for her assistance during my visits to the laboratory. Research has been supported by a grant from the National Geographic Society for investigations of deep-sea biology, and the operation of R/V PILLSBURY was supported by NSF grant GB-5776.

Class PELECYPODA
Family Nuculidae
Genus *Nuculana* Link, 1807

Type-Species.—*Arca rostrata* Chemnitz (= *Mya pernula* Müller), by monotypy. Recent, Europe.

Subgenus *Costelloleda* Hertlein & Strong, 1940

Type-Species.—*Nucula costellata* Sowerby, by original designation.

Nuculana (Costelloleda) costellata (Sowerby)
Fig. 12

Nucula costellata Sowerby, 1833, Proc. Zool. Soc. Lond., 1832: 198. "Hab. ad Panamam. Ten fathoms."
Leda (Leda) costellata (Sowerby), Dall, 1908, Bull. Mus. Comp. Zool. Harv., 43(6): 375.
Nuculana (Costelloleda) costellata (Sowerby), Hertlein & Strong, 1940, Zoologica, N.Y., 25(4): 398-399, pl. 2, fig. 10.

Remarks.—An elongated shell with a pale olivaceous periostracum, the surface underneath white and sculptured with raised, distant concentrics. Figured specimen measures: length 15.7 mm, height 5.8 mm, thickness of paired valves 3.5 mm.

Material Examined.—PILLSBURY Sta. P-488, Gulf of Panama, 8°13.1'N, 80°09.6'W, depth 17 m, 1-2 May 1967. UMML 28-2297.—PILLSBURY Sta. P-543, Gulf of Panama, 8°21.4'N, 78°45.5'W, depth 62-70 m, 7 May 1967. UMML 28-2727.

Genus Adrana H. & A. Adams, 1858

Type-Species.—Nucula (Adrana) lanceolata Lamarck, by subsequent designation: Stoliczka, 1871.

Adrana exoptata (Pilsbry & Lowe)
Figs. 2, 3

Leda (Adrana) exoptata Pilsbry & Lowe, 1932, Proc. Acad. Nat. Sci. Philadelphia, 84: 107, pl. 17, figs. 8, 9.
Adrana exoptata (Pilsbry & Lowe), Olsson, 1961, Panamic-Pacific Pelecypoda: 71, pl. 3, fig. 6.

Remarks.—The shell of this species is small, delicate, with a straight dorsal margin and weak to strong sculpture. A single paired specimen in the collection. Length 15.0 mm, height 10.0 mm.

Material Examined.—PILLSBURY Sta. P-488, Gulf of Panama, 8°13.1'N, 80°09.6'W, depth 17 m, 1-2 May 1967. UMML 28-2296.

Adrana marella (Hertlein & Strong)
Fig. 11

Nuculana (Costelloleda) marella Hertlein & Strong, 1940, Zoologica, 25(4): 399, 400, pl. 2, figs. 12, 13.

Remarks.—Shells of this species are fairly large, with fine concentric sculpture. Figured specimen, a right valve, measures: length 27.8 mm, height 10.7 mm, semidiameter 2.5 mm.

Material Examined.—PILLSBURY Sta. P-543, Gulf of Panama, 8°21.4'N,

FIGURES 1-12. Panamic Pelecypoda: 1, *Limopsis compressus* Dall; length without periostracum 36.5 mm, height 28.8 mm, UMML 28-2717, PILLSBURY Sta. P-526.—2, 3, *Adrana exoptata* (Pilsbry & Lowe); paired valves, length 15 mm, UMML 28-2296, PILLSBURY Sta. P-488.—4-6, *Tellina (Tellina) argis*, new species, holotype; 4, 5, right valve, exterior and interior, length 32.1 mm; 6, magnified view of surface to show details of sculpture; ARGOSY Sta. 38.— 7-10, *Chama corallina*, new species, holotype, paired valves; height 22.6 mm,

78°45.5'W, depth 62-70 m, 7 May 1967. UMML 28-2644, 6 specimens; also a fragment of a larger shell from the same station, UMML 28-2656.

Family Limopsidae

Genus *Limopsis* Sacco, 1827

Type-Species.—*Arca aurita* Brocchi, by monotypy. Upper Neogene of Italy.

Limopsis compressus Dall
Fig. 1

Limopsis compressus Dall, 1896, Proc. U.S. Nat. Museum, *18:* 16; 1908, Bull. Mus. Comp. Zool., *43*(6): 392, pl. 7, figs. 7, 8. (USS ALBATROSS Sta. 3382, Gulf of Panama, 1793 fathoms = 3279 meters, mud bottom, temperature 36°F. Type, USNM 122889.)

Description.—The shell is rather large, with the valves thin, compressed, obliquely ovate, the small, inconspicuous beaks closer to the anterior end; it is externally covered with a yellowish brown, pilose periostracum, the surface below white, polished, and marked with a faint reticulate pattern of concentrics and radials. Unusual for its large size.

Measurements.—Length exclusive of periostracum, 35 mm, height 28 mm, diameter 9.9 mm (an average specimen).

Material Examined.—PILLSBURY Sta. P-526, Gulf of Panama, 6°53'N, 79°27'W, depth 3193-3200 m, 5 May 1967. UMML 28-2717, 15 perfect specimens and one free valve.

Family Chamidae

Genus *Chama* Linné, 1758

Type-Species.—*Chama gryphoides* Linné, by subsequent designation: Schumacher, 1817. Mediterranean.

Chama corallina, new species
Figs. 7-10

Description.—The shell is small with rounded plump valves of a pinkish or coral red color, attached by a small area on the umbo of the left valve.

←

length 20 mm, PILLSBURY Sta. P-549.—11, *Adrana marella* (Hertlein & Strong); length 27.8 mm, height 10.7 mm, UMML 28-2644, PILLSBURY Sta. P-543.— 12, *Nuculana (Costelloleda) costellata* (Sowerby); length 15.7 mm, height 5.8 mm, UMML 28-2297, PILLSBURY Sta. P-488.

The left or attached valve is a little larger than the right, its umbonal section coiled so as to resemble a small, Cretaceous chamoid. Both valves are of nearly equal convexity. Surface sculpture consists of small, crowded, radial wrinkles and on an occasional specimen there are a few lines of small, radial, sharp spines. The internal edge of the valve margin shows two shell layers; the outer one, probably calcitic, is minutely wrinkled and pinkish, and the inner layer, also finely crenulated, is white. The hinge is normal, with a central, rugose cardinal tooth. External color is a pale coral red or pink, the interior white.

Measurements.—Height 22.6 mm, length 20.0 mm, diameter (closed valves) 18.6 mm. Holotype.

Holotype.—USNM No. 701157. PILLSBURY Sta. P-549, Gulf of Panama, off Punta Escondido, R.P., 7°59.5'N, 78°30.3'W, depth 55 m, 7 May 1967.

Paratypes.—An additional 37 specimens and 17 odd valves from the same station, UMML 28-2640.

Remarks.—This is a small, rounded globose species attached by the umbone of its left valve, generally to pebbles, and easily distinguished from other Panamic species by its shape and reddish color.

Family Tellinidae
Genus *Tellina* Linné, 1758

Type-Species.—*Tellina virgata* Linné, by subsequent designation: Lamarck, 1799. Indo-Pacific.

Tellina (Tellina) argis, new species
Figs. 4-6

Description.—The shell is broadly lanceolate, the posterior side shorter, flexed, the longer anterior end obliquely rounded at the end. The sculpture is pronounced, consisting of numerous flattened, concentric bands of ribbons over the anterior half of the disk, but which in the belt between the middle and posterior umbonal flexure rise into divided, erect lamellae. The posterior umbonal flexure in the right valve is a wide, elevated ridge extending from behind the beak to the posterior ventral margin where it forms a subtruncated point, and over which the concentric lamellae extend. The posterior dorsal slope is narrow and flat, and likewise sculptured with concentric lamellae in alternate strength. The interior is smooth, the adductor scars and pallial impressions are indistinct. Hinge of the right valve has two small cardinal teeth, the posterior one bifid, the lateral sockets are distant and long.

Measurements.—Length 32.1 mm, height 18.1 mm, diameter 3.0 mm; holotype.

Holotype.—USNM No. 701158. ARGOSY Sta. 38, south of Gorgona Island, Colombia, 2°39′N, 78°38′W, depth 91-101 m, 23 September 1961.

Remarks.—Only a single specimen, a right valve, is known of this interesting species, but its sculpture, unique amongst American tellinids, should serve to identify the species without difficulty.

Class GASTROPODA
Subclass *PROSOBRANCHIATA*
Superorder NEOGASTROPODA
Order TOXOGLOSSA

Family Terebridae

Genus *Terebra* Bruguière, 1789

Type-Species.—*Buccinum subulatum* Linné, by monotypy.

Remarks.—Several large Terebras occur in the Panamic-Pacific region and are separated as distinct species on the basis of shape, sculpture (mainly of the spire whorls), and color pattern. A single specimen taken by the ARGOSY differs from the named forms in many respects and is described below as a new species.

Terebra argosyia, new species
Figs. 28-30

Description.—The shell is fairly large, stout, nearly straight sided, tapering uniformly to a minute, needle-sharp apex. Whorls are numerous, about 20 on the type, with a tripartite sculpture most marked on the spire but persistent over the whole surface. The sculpture consists of a rather large, axially noded, convex cord bordering the suture and, below, a more or less depressed zone divided by a groove into two subsidiary bands. The columella has two sharp plaits. Color pattern on the spire whorls between the sutures consists of two bands of brown spots or flammules on a white or yellow base. On the body whorl, a fourth much larger and more heavily colored band encircles the base.

Measurements.—Length 71.4 mm, diameter 15.0 mm; holotype.

Holotype.—USNM No. 701160. ARGOSY Sta. 72, Isla la Plata, Ecuador; east side of island at anchorage in a depth of 37 m, taken by diving, October 7, 1961.

Remarks.—Represented in the collection by a single specimen.

Superfamily Conacea
Family Turridae *sensu lato*

The gastropods commonly known as turrids and generally grouped together in a single family "Turridae" now appear to be a mixture of many unrelated genera which are best differentiated amongst the Recent species on radular characters. It is to Thiele in his report on the Valdivan turrids (1925) and again in his Handbuch der Systematischen Weichtierkunde (1928) that we owe the first indication of this revelation, and the accumulation of radular information since obtained fully supports this contention. For convenience of treatment in this review, as well as for offering a preliminary classification, the turrids are divided into two major groups: first, the true turrids (the Turridae proper) which possess a radular ribbon which in its most complete form has five distinct teeth in each transverse row (the *Drillia* complex) or, as in most others (*Turris, Polystira, Hormospira,* etc.), has the lateral and marginal teeth more or less combined so as to appear to resemble a single tooth while the central teeth may be fully developed, vestigial, or lacking. In the ribboned turrids, the radular organ lies within the proboscis, connected with the buccal mass, and it resembles the radula of other ribboned gastropods so closely that its operation must be that of a rasping device. The question which remains is whether or not the teeth of the ribboned turrids are able to deliver a toxic dosage more potent than a mere surface irritant.

The second group comprises the toxoglossate turrids which are provided with needlelike, or dart- and arrow-shaped teeth similar to those of the cones. Although the anatomy of most turrids is unknown, these teeth apparently are developed in a radular sack within the body proper behind the proboscis in the same manner as in the cones. The individual radular teeth of these turrids are often indistinguishable from those of the cones and many are provided with hold-fast barbs and blades on the sides. These snails are truly toxoglossates, their teeth designed to deliver a toxic discharge directly into the body of a living prey, and the difference between these two major groups—those with a rasping ribbon and those without—is too deep for both to belong to a single family. The case is, however, not quite so simple because of our general lack of supporting anatomical knowledge, and there are several known instances of radular dimorphism amongst some species. Until these contradictions have been resolved, the hasty naming and reassignment of genera on radular characters alone should be avoided.

Family Turridae *sensu stricto*
(The "ribboned turrids")

The ribboned turrids are fairly well represented in the PILLSBURY collections. Most species were obtained in relatively small numbers, except for those of *Polystira* which, at several stations, were taken in bulk and demonstrate that species of this genus are amongst the commonest of medium-sized mollusks in offshore waters.

Genus *Drillia* Gray, 1838

Type-Species.—*Drillia umbilicata* Gray, by subsequent designation: Gray, 1847. West Africa.

Remarks.—Species of the *Drillia* complex are sparingly represented in the PILLSBURY collection and nominally referred to several genera on the basis of shell characters. The species will be listed in another report. On radular characters, all Drillias are readily recognizable by a large, rounded, gear-shaped lateral tooth, while the central tooth is small, nearly vestigial. The general radular pattern is: 1.1.1.1.1, or quinqueserial with 5 distinct teeth.

A figure of the radula of *Drillia rosolina* Marrat (Fig. 87), from West Africa, is introduced for comparison and interpretation of other ribboned turrids.

Genus *Hormospira* Berry, 1958
= *Tiariturris* Berry, 1958; type-species, *T. spectabilis* Berry

Type-Species.—*Pleurotoma maculosa* Sowerby, by original designation.

Description.—The shell is fairly large, elongate-fusiform, with a high, attentuated spire and wide anterior canal. The protoconch is small and consists of about three whorls; it is directly followed by the nepionic whorls sculptured with axial riblets and fine spirals. The mature whorls are shouldered by the edge of the subsutural fasciole. The sculpture is produced mainly by knobbed axial riblets, strongest on the spire whorls and fading out on the later ones, their surface becoming smooth except for spiral threads along the anterior canal. The margin of the lip is thin, sinuated in the middle and notched above by a wide, deep, U-shaped anal sinus forming a deep fasciole between the shoulder and the suture, smooth except for the growth lines of the sinus. The anterior canal is large, long, and straight, or with a slight twist at its end, the siphonal canal not forming a fasciole. The radula is a large, falsely triserial ribbon in which the central tooth is a large, rectangular plate with a single, solid cusp in the middle and the lateral and marginal teeth are double, elongated, the marginal smaller, loosely overlying the lateral. The formula may be written 1.1.1.1.1.

Remarks.—*Hormospira maculosa* Sowerby is a common offshore species in the Gulf of Panama and was trawled in numbers at several stations in depths of 18 to 117 meters. Fresh shells have a dark, olivaceous brown periostracum; the surface underneath is white, blotched with brown, especially on the spire. The radula is illustrated in Figure 88.

Genus *Steiraxis* Dall, 1895

Type-Species.—*Pleurotoma* (*Steiraxis*) *aulaca* Dall, by monotypy.

Description.—"Shell resembling *Irenosyrinx*, but with stronger sculpture and with a paucispiral operculum" (Dall, 1908). To this brief description may be added: the radula is a ribbon showing a triserial pattern, the central tooth small, the two elements of the lateral teeth overlapping each other (compound).

Steiraxis aulaca Dall
Figs. 20, 90

Pleurotoma (*Steiraxis*) *aulaca* Dall, 1896, Proc. U. S. Nat. Mus., *18:* 14 (Fish Commission station 3415, in 1879 fathoms = 3437 meters, off Acapulco, Mexico; USNM 123099).
Steiraxis aulaca, Dall, 1908, Bull. Mus. Comp. Zool., *43*(6): 273, 274; pl. 2, fig. 5.

Description.—During the PILLSBURY cruise in the Gulf of Panama, three living specimens of this turrid were obtained at Station 526 from a depth of 3193-3200 meters, associated with several other deep-water mollusks. The soft parts of one specimen yielded the radula (see Fig. 90). The shell is quite large, thin, white, the spire whorls deeply corroded, revealing the smooth surface of the inner layer. The sculpture is confined to the relatively thin outer layer and consists of small, close-set, sharp, spiral cords, those in the middle grouped together to form a stout, composite peripheral keel. The margin of the lip is thin, serrated by the spiral cords, widely sinuated to form a wide anal sinus between the peripheral cord and the suture. The radula is a ribbon with a false triserial pattern, the central tooth being small, triangular in shape, thin and with a medial cusp. The side teeth are compound, the lateral tooth overlain by the marginal.

Measurements.—The PILLSBURY specimens measure as follows: length 92.0 mm, diameter 38.0 mm; length 86.7 mm, diameter 37.4 mm; length 88.4 mm, diameter 37.0 mm (figured specimen).

Material Examined.—PILLSBURY Sta. P-526, Gulf of Panama, 6°53'N, 79°27'W, 3193-3200 m, 5 May 1967. UMML 30-5484, three specimens.

Genus *Knefastia* Dall, 1919

Type-Species.—*Pleurotoma olivacea* Sowerby, 1834 (= *P. funiculata* Val., *in* Kiener, 1839), by original designation. Recent, Panamic faunal province.

Description.—The shell is solid, fasciolarid in shape but with a straight, smooth, pillared anterior canal. The whorls are shouldered by the lower edge of a deeply concave, subsutural fasciole marked with the bowed trace growth lines of a wide, U-shaped anal sinus. The sculpture is fusoid, formed by foldlike riblets, heaviest along the shoulder and overridden by spiral cords. The radula is a ribbon with an unequal-armed, wishbone-shaped lateral tooth, a smaller, loose marginal, and a minute, vestigial central tooth.

Remarks.—Several *Fasciolaria*-like turrids have been referred to *Knefastia*, but until their radular characteristics are known and fully illustrated, their generic position is uncertain. Judging by Powell's figure of the radula of *K. funiculata* Kiener (a possible genotype), *Knefastia* appears to be related to *Turricula*, but further radular investigation would be desirable.

Knefastia? pilsbryi (Lowe)
Figs. 16, 17, 89

Clathrodrillia pilsbryi Lowe, 1935, Trans. San Diego Soc. Nat. Hist., 8(6): 23, 24; pl. 4, fig. 2 (Punta Penasco, Sonora, dredged, 10 fathoms = 18 m).

Description.—The shell is quite large, solid, fusiform in shape and sculpture, with an elevated, attenuated spire of eight or more whorls, its height about half again that of the aperture, the anterior canal stout, straight, and weakly recurved at the end. The subsutural fasciole is a deeply constricted zone between the shoulder and the suture; unmarked except for faint spiral threads and the bowed growth lines of the sinus. The dominant sculpture is axial, formed by knobbed riblets around the middle and along the shoulder (eight on the body whorl), absent from the fasciole above and nearly so across the base and canal, the whole overridden by spiral cords in a banded fashion. Margin of lip is sinuated, not thickened, widely flaring in the middle and curving inward above into the wide, deep, anal sinus in the middle of the fasciole separated from the suture by a space developing a callus which spreads down along the parietal wall to the tip of the canal.

The radula is a falsely biserial ribbon, the lateral and marginal teeth of simple, slablike shape, pointed, the smaller marginals inconspicuously overlying the laterals, no central teeth, the middle zone of the ribbon open and empty (see Fig. 89).

Measurements.—Length 53.4 mm, diameter 18.6 mm.

Material Examined.—PILLSBURY Sta. P-502, Gulf of Panama, 7°40'N, 79°50.5'W, depth 79-77 m, 2 May 1967. UMML 30-5293, 6 specimens.
—PILLSBURY Sta. P-513, Gulf of Panama, 7°39.5'N, 79°40.7'W, depth 117-108 m, 4 May 1967. UMML 30-5387, 2 specimens.

Remarks.—The presence of this species in the Gulf of Panama represents an extension of its range south from the Gulf of California. Its generic reference to *Knefastia* is tentative.

The Toxoglossate Turrids

The toxoglossate turrids comprise a large number of forms assigned to different genera whose true status is dependent on radular assessment. Amongst the PILLSBURY forms, the following are of special interest.

Family Brachytomidae Thiele, 1929

Thiele's definition of the family (used originally as a subfamily) is as follows: An operculum is present. Radula without a basal membrane, with more or less elongated, sharp, fluted teeth. The type-genus is *Brachytoma* Swainson, 1840.

Amongst the turrids in the PILLSBURY collection from the Gulf of Panama are several specimens of *Strombinoturris crockeri* Hertlein & Strong, described originally by its authors as a new genus and species and referred to the Columbellidae because of a general likeness of its outer lip to *Strombina*, although notched by an anal sinus. Mr. Paul Leviten dissected a specimen of *Strombinoturris* and proved it to be a turrid, later more fully confirmed by the toxoglossate radula. Further studies at the British Museum (Natural History) showed that *Strombinoturris crockeri* is the long misunderstood *Pleurotoma stromboides* Sowerby, 1832, described without mention of locality. The type-specimen of *Pleurotoma stromboides* in the British Museum agrees perfectly with its figure in the *Genera of Shells* (vol. 2, pl. 215, pl. 4). It is equally certain that Petit's *Pleurotoma sumatrensis*, although described as from Sumatra, also represents the Panamic species. It is well to note that the other species described by Petit in the same paper with *P. sumatrensis* are all from America. This situation confirms Reeve's

---

FIGURES 13-27. Panamic Gastropoda: 13, *Brachytoma stromboides* (Sowerby); length 39.1 mm, UMML 30-5375, PILLSBURY Sta. P-530.—14, 15, *Glyphostoma bayeri*, new species; 14, a dead specimen, length 24.3 mm, paratype, PILLSBURY Sta. P-529; 15, fresh specimen with color, length 25.7 mm, holotype, PILLSBURY Sta. P-521.—16, 17, *Knefastia? pilsbryi* (Lowe); 16, length 52.5 mm, UMML 30-5293, PILLSBURY Sta. P-502; 17, length 58.8 mm, UMML 30-5293, PILLSBURY Sta. P-502.—18, *Phymorhynchus speciosus*, new species; length 39.7 mm, diameter 18.4 mm, holotype, PILLSBURY Sta. P-526.—19, 24, 25, *Bayerius fragilissimus* (Dall); 19, length 31 mm; 24, length 29.3 mm, both PILLSBURY Sta. P-526; 25, length 21 mm, holotype, USNM No. 123007.—20, *Steiraxis aulaca* Dall; dorsal aspect of an imperfect shell, length 88.4 mm, UMML 30-5484, PILLSBURY Sta. P-526.—21, *Carinodrillia dariena*, new species; length 34.7 mm, holotype, PILLSBURY Sta. P-493.—22, *Carinodrillia alcestris* (Dall); length 43.5 mm, UMML 30-5423, PILLSBURY Sta. P-553.—23, *Mangilia enora* Dall; length 23 mm, PILLSBURY Sta. P-526.—26, 27, *Carinodrillia bicarinata* (Shasky); length 46.6 mm, diameter 16.5 mm; front view showing pattern, side view of same specimen whitened, showing anal sinus, UMML 30-5292, PILLSBURY Sta. P-502.

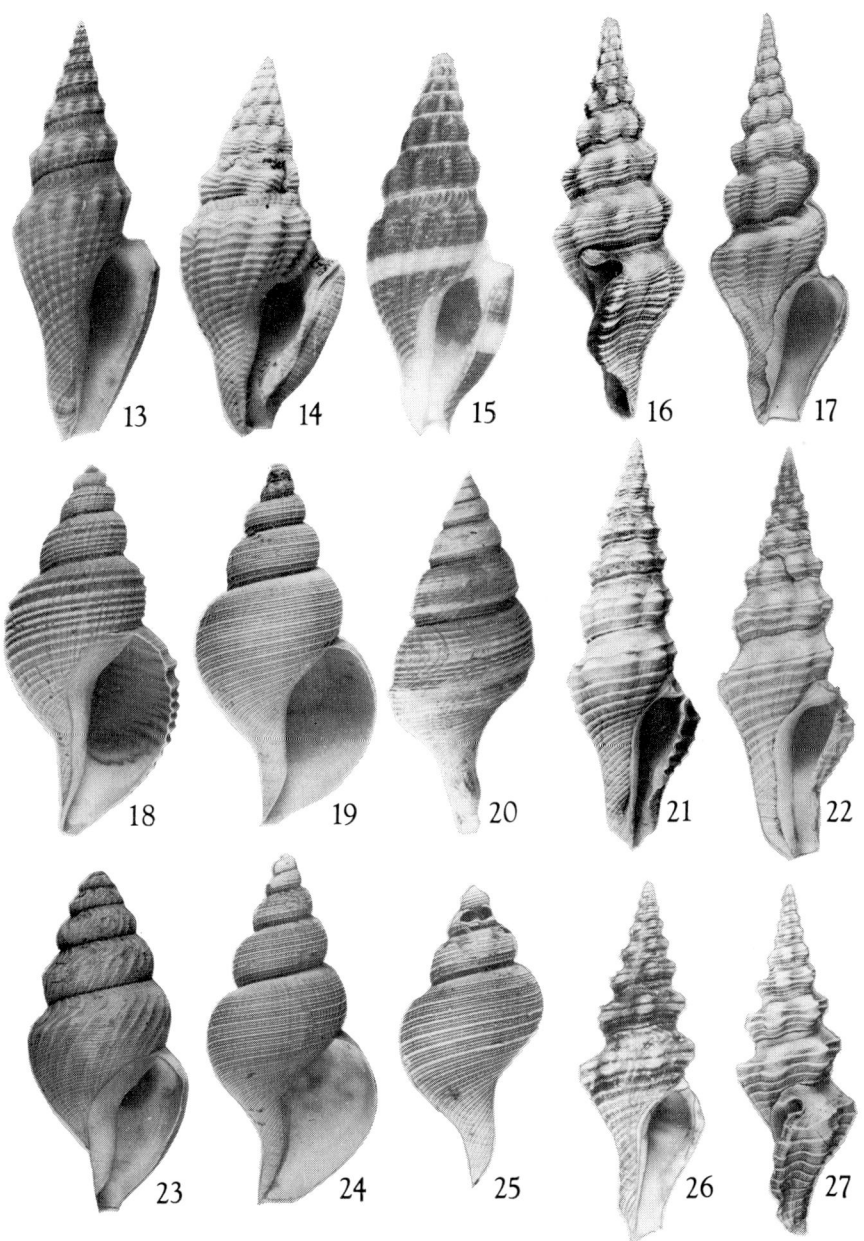

record and identification of *Pleurotoma stromboides* as a Panamic species on the basis that the illustrated specimen was collected by Hinds in the Bay of Panama.

Genus *Brachytoma* Swainson, 1840

= *Strombinoturris* Hertlein & Strong, 1951, type-species, *S. crockeri* Hertlein & Strong

Type-Species.—*Pleurotoma strombiformis* Sowerby, 1839 (= *Pleurotoma stromboides* Sowerby, 1832, *Genera of Shells*, fig. 4) by subsequent designation: Herrmannsen, 1846.

Description.—The shell is strombiform, with high, pointed spire and straight anterior canal, the outer lip margin expanded and thickened by a varical rib on the back. The anal sinus is a deep, U-shaped notch in the upper part of the margin, forming a subsutural fasciole marked mainly with the growth lines of the sinus and often with a frilled cord at the suture. Aperture is smooth or without denticles on the sides. Sculpture of axial riblets begins at the shoulder and extends down over the base, overridden by spirals. Radula is toxoglossate, with large, slender, needle-like teeth.

Remarks.—*Brachytoma* is a glyphostomid in shell and radular characters and differs from *Glyphostoma* mainly in having the aperture smooth on both sides and with less elaborate sculpture. The sutural frill so characteristic of typical *Glyphostoma* is poorly developed in *Brachytoma stromboides* and wholly lacking in some specimens.

Brachytoma stromboides (Sowerby)
Figs. 13, 91

Pleurotoma stromboides Sowerby, 1832, *Genera of Shells*, vol. 2: pl. 215, fig. 4. (No description or mention of locality; type-specimen in the British Museum [Natural History], register number L. 196-991.)

Pleurotoma stromboides Sowerby, Reeve, 1843, Conch. Icon., *1:* sp. 71, pl. 1, fig. 71. (Hab. Bay of Panama [found in mud at the depth of seven fathoms]; Hinds.)

Pleurotoma sumatrensis Petit de la Saussaye, 1852, Journ. Conchyliol., *3:* 55, pl. 2, fig. 2. (The locality given as off Sumatra is probably incorrect.)

Strombinoturris crockeri Hertlein & Strong, 1951, Zoologica, *36*(2): 84, pl. 1, fig. 9. (Type, Arena Bank, Gulf of California.)

Remarks.—Shape and other characteristics of the shell are as figured. The radula is toxoglossate, with hollow-shafted, needle-like teeth similar to those of *Glyphostoma*. In a specimen from PILLSBURY Sta. P-529, the radular sack was found to be quite large and contained 63 teeth averaging 1.9 mm in length. An individual tooth resembles a transparent, slender needle with a sharp point and with a small hook or barb directly behind.

The shaft is hollow, as indicated by a line of bubbles within. The wall is solid, without an evident lateral seam.

Material Examined.—PILLSBURY Sta. P-513, Gulf of Panama ENE of Cabo Mala, 7°39.5′N, 79°40.7′W, depth 117-108 m, 4 May 1967, UMML 30-5388, 1 specimen.—PILLSBURY Sta. P-515, Gulf of Panama NE of Cabo Mala, 8°00.4′N, 79°40.8′W, depth 79-77 m, 4 May 1967, UMML 30-5446, 2 specimens.—PILLSBURY Sta. P-529, Gulf of Panama, S of Isla de San José, 8°00.7′N, 79°11.8′W, depth 84 m, 6 May 1967, UMML 30-5344, 1 specimen.—PILLSBURY Sta. P-530, Gulf of Panama, W of Isla de San José, 8°15.3′N, 79°10.6′W, depth 66-71 m, 6 May 1967, UMML 30-5375, 2 specimens.—PILLSBURY Sta. P-531, Gulf of Panama, W of Isla Pedro Gonzales, 8°25.5′N, 79°10.7′W, depth 57-64 m, 6 May 1967, UMML 30-5332, 11 specimens.

Genus *Glyphostoma* Gabb, 1873

Type-Species.—*G. dentiferum* Gabb, by monotypy. Miocene of Santo Domingo.

Description.—The shell is strombiform, fusiform, with high spire and straight anterior canal, generally finely sculptured and with an expanded lip. In typical species, the lip is widely expanded, thickened by a varical rib on the back, its forward face flat, its inner surface smooth or lirated. The subsutural fasciole is a wide, concave space marked with the deep growth lines of the retral sinus and bordered by a band of small, axial frills at the suture. The protoconch is a small, turbinate coil of two or more whorls, smooth or carinate. (See Olsson, 1964:105.)

Remarks.—The genus "Glyphostoma" includes a large number of species, often with elaborate cancellated sculpture, the majority of which are fossil (Miocene). Many have coarse denticulation on both sides of the aperture; in others this is lacking or of variable strength. About 12 species have been described from the eastern Pacific, but most are from deep water and are not members of the Panamic-Pacific fauna in a strict sense. All species are rare and sparingly represented in museum collections.

Glyphostoma bayeri, new species
Figs. 14, 15

Description.—The shell is of medium size, relatively stout, with a high spire and large rounded body whorl, the anterior canal of medium length, straight except at the tip, which is recurved. The subsutural fasciole is deep and narrow, strongly sculptured with small axial riblets crossed by three spiral threads (strongest in their intervals) and which cover the entire space. The sculpture is fusoid, formed by strong, rounded axial riblets (11 on the body whorl) overridden by regular primary spiral cords of which there are about 16 on the body from the edge of the subsutural fasciole to the middle of the base, with smaller and more crowded spirals along the back of the anterior canal; about five spiral cords cross the axials on the penultimate whorl between the fasciole and lower suture. The outer lip is expanded, stromboid, edged by a large varical rib, flattened in front and bordered behind by a deeply impressed groove, the anal sinus wide and deep. The apertural denticles in the type are weak and lie only in the middle on the outer lip and on the columella on the inner. Color of fresh specimens is brown with a white band in the middle.

Measurements.—Length 25.7 mm, diameter 10.7 mm, holotype.

Holotype.—USNM No. 701211. PILLSBURY Sta. P-521, Gulf of Panama, 7°48.3'N, 79°35.1'W, depth 119 m, 4 May 1967.

Paratypes.—UMML 30-5352. PILLSBURY Sta. P-529, Gulf of Panama, 8°00.7'N, 79°11.8'W, depth 84 m, 6 May 1967; two specimens.

Remarks.—The holotype is a dead shell, fresh, white with a trace of color. The species resembles *G. dentiferum* Gabb of the Miocene of Santo Domingo, and also *G. gabbi* Dall of the Caribbean, the most obvious differences from those species being its wider body whorl, weaker lip denticles, and coloration.

Genus *Cruziturricula* Marks, 1951

Type-Species.—*Turricula* (*Pleurofusia*) *cruziana* Olsson, by original designation. A Miocene species from northwestern Peru.

Description.—The shell is fusiform, with an elevated, tapered spire of many whorls, the anterior canal straight and narrow, the sculpture dominantly spiral. The apical whorls form a narrowly attenuated spire, the protoconch itself being a small bulbous cap, smooth and slightly carinated, the following whorls with bowed axials above the middle and spirals below. The postnuclear and mature whorls are peripherally shouldered, weakly undulated by very small nodes; the space between the shoulder and suture is quite high, with a flatly sloping or somewhat depressed surface containing an inconspicuously defined subsutural fasciole in its upper half. The margin of the outer lip is thin, widely sinuated forward in the middle and retractive above into a deep anal notch, bordering the suture, its lower limb well above the shoulder, large, straight, its upper shorter limb confluent with the suture. Texture relatively thin, the sculpture composed of sharp primary spiral cords principally developed along the peripheral shoulder and across the base and onto the canal, with finer spiral threads in their intervals. The radula of the living species (*C. panthea* Dall) is toxoglossate and consists of numerous, small, dagger-shaped teeth, the upper half narrow and sharply pointed at the end and with a hooked barb on one side and a blade and hook on the opposite; lower half of the tooth shaft is wide with a knuckle-shaped base and spur.

Remarks.—This genus was proposed by J. Marks with *Turricula cruziana* Olsson, a Miocene species from northwestern Peru, as type. The lineage is well represented amongst the fossils from the Eocene onward, its Recent representative being *C. panthea* (Dall) from the Gulf of Panama. Their shells are recognized in general by their fusoid shape with high spire and straight anterior canal which, together with a dominantly spiral sculpture, gives to the group a general resemblance to *Polystira*, but with a subsutural sinus. They differ from the *Turricula* complex by their toxoglossate radula (Fig. 93).

Cruziturricula panthea (Dall)
Fig. 93

Turricula (Surcula) panthea Dall, 1919, Proc. U. S. Nat. Mus., 56(2288): 4, pl. 1, fig. 6 (not fig. 5 as given on page 4 or in the explanation of plate). (ALBATROSS Sta. 2795, in Panama Bay, in 33 fathoms = 60 meters. USNM 212348.)

Remarks.—This is a fairly common species in the Gulf of Panama and was collected by the PILLSBURY at several stations in depths from 53 to 119 meters. The species is easily recognized by its fusoid shape, high spire, and straight, narrow anterior canal; the primary sculpture is spiral, the axials reduced to oblique nodes at the shoulder. The space between the shoulder and the suture is wide, sloping or slightly depressed, the anal sinus deep, with a rounded base, its short upper limb confluent with the suture. Color white or cream with brown flecks over the shoulder nodes.

In various works (Olsson, Keen, and others), this species has generally been cited under the name *Turricula lavinia* Dall, due to an unfortunate reversal of names in the text and in the explanation of the plate. This situation was pointed out to me recently by Dr. J. H. McLean of the Los Angeles County Museum of Natural History. Most workers everywhere identify their specimens by the published figures and, if the agreement is close, accept such identification without question. Nomenclatural rules, however, require that the type-specimen itself constitute the decisive factor in these cases.

Pleurotoma arcuata Reeve, 1843 (Conch. Icon., vol. 1, *Pleurotoma*, sp. 15, pl. 3, fig. 15) described from the coast of Veragua, Central America, is probably this species, but its figure is poor and the type (if preserved at the British Museum) should be examined to determine its relationship with *panthea*.

Material Examined.—PILLSBURY Sta. P-493, 7°39.5'N, 80°00.7'W, depth 46 m, 2 May 1967, UMML 30-5305, 2 dead specimens.—PILLSBURY Sta. P-498, 8°00'N, 79°49.8'W, depth 59 m, 2 May 1967, UMML 30-5418, 1 specimen.—PILLSBURY Sta. P-500, 7°59.7'N, 79°49.7'W, depth 53 m, 2 May 1967, UMML 30-5360, 14 specimens.—PILLSBURY Sta. P-502, 7°40'N, 79°50.5'W, depth 79-77 m, 2 May 1967, UMML 30-5290, 1 specimen.—PILLSBURY Sta. P-521, 7°48.3'N, 79°35.1'W, depth 119 m, 4 May 1967, UMML 30-5474, 4 dead specimens.—PILLSBURY Sta. P-530, 8°15.3'N, 79°10.6'W, depth 66-71 m, 6 May 1967, UMML 30-5377, 2 dead specimens.—PILLSBURY Sta. P-531, 8°25.5'N, 79°10.7'W, depth 57-64 m, 6 May 1967, UMML 30-5335, 11 specimens.—PILLSBURY Sta. P-543, 8°21.4'N, 78°45.5'W, depth 62-70 m, 7 May 1967, UMML 30-5463, 3 dead specimens.—PILLSBURY Sta. P-550, 7°59.4'N, 78°40.3'W, depth 64-60 m, 7 May 1967, UMML 30-5317, 1 specimen.—PILLSBURY Sta. P-553, 7°59.2'N, 79°01'W, depth 99 m, 7 May 1967, UMML 30-5422, 4 specimens.—PILLSBURY Sta. P-555, 7°50.7'N, 79°00.3'W, depth 68 m, 7 May 1967, UMML 30-5408, 1 specimen.

Genus *Carinodrillia* Dall, 1919

Type-Species.—*Clathrodrillia (Carinodrillia) halis* Dall, by original designation. Recent, Lower California.

Description.—The shell is fusiform, the spire longer than the aperture, the whorls have a rounded or angled shoulder, the color is usually dark brown. The protoconch is small, turbinate, of three whorls, the first smooth, the second slightly keeled, the third with bowed axials, the change to the nepionic gradual. The subsutural fasciole is wide and deeply excavated, appressed against the suture and often with a subsutural cord, its lower surface smooth except for the growth lines of the sinus and sometimes weak spirals. The anal sinus is a deep U-shaped notch in the upper part of the apertural lip and, in the mature stage, its edges are raised or reflexed. The sculpture is fusoid, with axial riblets strongest in the peripheral zone, overridden by spiral cords. The aperture is clavate, its outer lip margin bowed or straight in the middle, thickened on the back by the last axial riblet. The anterior canal is stout, quite long, and straight except at the end, which is twisted to the right. The radula is toxoglossate, with pointed awl-shaped teeth with a barb on one or both sides (see Fig. 92, *C. dariena*; and Fig. 94, *C. duplicata*).

Remarks.—*Carinodrillia* is herein considered a toxoglossate genus as demonstrated by its radula, extracted from several species assigned to it on shell characters. McLean has recently informed me that *C. halis* (Dall), its type-species, has a ribboned radula. The resemblance of *C. halis* to such species as *C. alcestris* is so close, approaching identity, that their placement in different genera as well as families is hardly warranted on basis of present information.

Carinodrillia dariena, new species
Figs. 21, 92

Description.—The shell is of medium size, solid, slender fusiform, the spire nearly twice the height of the aperture, its whorls about 11 in number, roundly angulated at the shoulder and decreasing in size upward, the apical end very small and sharp. The anterior canal is stout and quite wide, its end twisted toward the right and enclosing a minute umbilical chink in the middle, the siphonal canal forming a low, fasciolar fold. The sculpture is fusoid, produced by axial riblets overridden by spiral cords, the general pattern shown in Figure 21. On the body whorl there are about 8 riblets, largest along the shoulder where they are sharply noded by the overriding spiral cords; above, the axials barely modulate the surface of the subsutural fasciole, while below they weaken out over the base toward the anterior canal. Spiral cords are sharp and narrow above, set between wide intervals, and smaller and closer together on the base and around the anterior canal, as illustrated in the figure. The aperture is elongate, wider above, the outer lip margin sharp and serrated by the ends of the spiral sculpture, and with the anal sinus drilloid in shape between the shoulder and suture. The terminal axial riblet on the back of the lip is somewhat enlarged. The radula is toxoglossate as shown in Figure 92.

Measurements.—Length 34.7 mm, diameter 10.4 mm, holotype.

Holotype.—PILLSBURY Sta. P-493, Bay of Panama, N of Isla Iguana, 7°39.5'N, 80°00.7'W, depth 37-33 m, 2 May 1967.

Carinodrillia alcestris (Dall)
Fig. 22

Clathrodrillia (Carinodrillia) alcestris Dall, 1919, Proc. U.S. Nat. Mus., 56 (2288): 18, pl. 5, fig. 6. (Off Guaymas in the Gulf of California in 20 fathoms. USNM 212354.)

Measurements.—Length 43.5 mm, diameter 15.3 mm, UMML 30-5423, from PILLSBURY Sta. P-553.

Material Examined.—PILLSBURY Sta. P-502, 7°40'N, 79°50.5'W, 79-77 m, 2 May 1967, UMML 30-5295, 7 specimens.—PILLSBURY Sta. P-519, 7°59.4'N, 79°34.3'W, depth 88 m, 4 May 1967, UMML 30-5327, 1 specimen.—PILLSBURY Sta. P-529, 8°00.7'N, 79°11.8'W, depth 84 m, 6 May 1967, UMML 30-5343, 4 specimens.—PILLSBURY Sta. P-530, 8°15.3'N, 79°10.6'W, depth 66-71 m, 6 May 1967, UMML 30-5380, 1 specimen.—

PILLSBURY Sta. P-553, 7°59.2'N, 79°01'W, depth 99 m, 7 May 1967, UMML 30-5423, 1 specimen.

Remarks.—Similar to *Carinodrillia duplicata* (Sowerby) but stouter and with a more angular form.

Several specimens agreeing well with Dall's figure of this species are in the PILLSBURY collection. Most of them were taken alive and have dark brown color. The radula is toxoglossate. *Carinodrillia halis* (Dall) appears to be a closely similar species with no obvious differences except that of a shorter canal. The type-specimen of *halis* in the National Museum of Natural History at Washington is a small, immature shell of white color, with partly fractured lip. It measures about 20 mm (19.7 mm according to my measurements). It was taken off La Paz. Two other immature, dead specimens of *halis* were secured by the ALBATROSS in the Gulf of California at a nearby station.

Carinodrillia bicarinata (Shasky)
Figs. 26, 27

Clathrodrillia (*Carinodrillia*) *bicarinata* Shasky, 1961, Veliger 4(1): 21, pl. 4, fig. 10. Various stations in the Gulf of California.

Description.—The shell is fusiform, with an elevated spire of ten or more whorls, the apical ones usually destroyed, and a somewhat shorter, straight anterior canal. Whorls are strongly shouldered and the subsutural fasciole is wide and deeply contracted, its surface plain except for fine spiral threads and faint, bowed growth lines of the anal sinus in the middle. The sculpture is dominated by a series of large, knobbed axial riblets (about seven on the body whorl) strongest along the shoulder and there overridden by two coarse, spiral cords so as to appear divided. Other primary spiral cords are straight, extending across the base and along the anterior canal in diminishing strength; these spiral cords are ridged, widely separated, their intervals bearing three or more finer secondary threads. The outer lip is thin, sinuous, flaring in the middle, the anal sinus in the contracted fasciole above, the notch deep, U-shaped, with contracted lips. Color of surface beneath is an ashy brown, the periostracum is a creamy white with brown stripes in the spiral intervals. The radula is toxoglossate, with needle-shaped teeth having a hooked barb directly behind the sharp point. Length of an individual tooth, 0.33 mm.

Measurements.—Length 46.6 mm, diameter 16.5 mm; length 52.2 mm, diameter 1.57 mm; PILLSBURY Sta. P-502. Length 39.3 mm, diameter 13.0 mm; PILLSBURY Sta. P-529.

Material Examined.—PILLSBURY Sta. P-502, 7°40'N, 79°50.5'W, depth 79-77 m, 2 May 1967, UMML 30-5292, 4 specimens.—PILLSBURY Sta.

P-529, 8°00.7′N, 79°11.8′W, depth 84 mm, 6 May 1967, UMML 30-5340, 12 specimens.

Remarks.—An interesting species of fusoid shape and characteristic markings. It bears no close resemblance to other Panamic species of *Carinodrilla*. The toxoglossate radula is high distinctive.

POSITION UNCERTAIN

Mangilia enora Dall
Fig. 23

Mangilia enora Dall, 1908, Bull. Mus. Comp. Zool., *43*(6): 286, pl. 4, fig. 6. (Off the coast of Ecuador, in 1132 fathoms = 2070 meters.)

Remarks.—A small, stubby shell with fine sculpture. The specimen, although containing the soft parts, did not yield a radula. The specimen measures: length 23.0 mm, diameter 11.3 mm.

Material Examined.—PILLSBURY Sta. P-526, 6°53′N, 79°27′W, depth 3193-3200 meters, 5 May 1967, UMML 30-7929, 1 specimen.

Order RACHIGLOSSA *sensu lato*

The term "Rachiglossa" is used herein in the emended, expanded sense of Troschel, in which the radular pattern is triserial and expressed in formula as 1.1.1 or 0.1.0. In the true Rachiglossa, as defined by Gray for such families as the Marginellidae and the Volutidae, only the central or rachidian tooth is commonly present, the laterals being retained fully only in a few genera such as *Volutocorbis*, lacking or vestigial in others.

Superfamily Olivacea
Family Olividae

The Olives are represented in the ARGOSY and PILLSBURY collections by only two species in the form of dead shells, most of them bleached white, only a few with good color. *Oliva kaleontina* Duclos was taken off Isla la Plata in Ecuador by the ARGOSY. A larger *Oliva* was trawled by both the ARGOSY and PILLSBURY at several stations in depths ranging from 29 to 77 meters. This species has not been precisely identified and may eventually require a separate name. It differs from *Oliva polpasta* Duclos, the common intertidal olive in Panama and southward, by its narrower form, richer color of mahogany brown (instead of an olivaceous shade), and especially by having a marked labial band shown by some specimens. A pronounced band (generally uncolored in Recent shells) is commonly developed on the outer side of the lip at maturity in many species, both

Recent and fossil, but has not previously been observed amongst Panamic forms. It is especially well marked in *Oliva waltoniana* Gardner of the Shoal River Miocene of Florida and in the Recent Caribbean *Oliva oblonga* Marrat, but its value as a systematic character needs additional investigation. For the sake of record, a specimen of the unidentified Panamanian shell is illustrated in Figure 35 (UMML 30-5177, Sta. P-553, 7°59.2'N, 79°01'W, depth 99 m). Specimens were taken also at Sta. P-493 (7°39.4' N, 80°00.7'W, depth 37-33 m; UMML 30-3905) and Sta. P-550 (8°00.7'N, 78°40.3'W, depth 64-60 m; UMML 30-5154).

Superfamily Fasciolariacea
Family Fasciolariidae
Genus *Latirus* Montfort, 1810

Type-Species.—*L. aurantiacus* Montfort (=*Murex filosus* Lamarck =*gibbulus* Gmelin), by original designation. Recent, Australia.

Latirus tumens Carpenter
Fig. 98

Latirus tumens Carpenter, 1856, Proc. Zool. Soc. Lond., pt. 24: 166. ("Hab. In Sinu Panamensi; legit Bridges. Sp. un. in Mus. Cuming.")

Remarks.—An excellent specimen of this, the rarest of the Panamic species of *Latirus,* was collected alive by the ARGOSY party at Isla la Plata, and extends its known range from Panama southward to Ecuador. The specimen measures as follows: length 88.3 mm, diameter 45.2 mm.

The radula extracted from the specimen is illustrated in Figure 98.

Superfamily Buccinacea
Family Buccinidae

The buccinids comprise a complex of marine genera whose typical members are mostly northern, but include many southern forms whose family kinship must be established by the radula. The radula is a triserial ribbon of which the central teeth are tricuspid (multicuspid in the nassarids), the lateral teeth conspicuously larger, more massive and with a prominent fanglike cusp at each end, and often smaller, accessory cusps in between. Characters of both shell and radula must be considered in the differentiation of genera amongst the Recent species.

Genus *Engina* Gray, 1839

Type-Species.—*Engina zonata* Gray, by subsequent designation: Gray, 1847.

Engina macleani, new species
Fig. 42

Description.—The shell is small, fusiform, with the spire slightly longer than the aperture. Whorls of the spire are about five (incomplete), the upper two being encrusted, forming a blunt apex. The general sculpture is fusoid, produced by short, stout axial riblets (about ten on the body whorl) crossed by a series of primary spiral cords between wider intervals, rising into link-shaped nodes over the axials and straight, narrow cords across the intervals; on the body whorl there are about 13 spiral cords over the rounded shoulder and across the base to the canal, together with a prominent subsutural cord, and a single fine secondary spiral occurs in each interval, most marked in the shoulder zone; on the spire whorls, there are three prominent spiral cords in addition to the subsutural one. The anterior canal is fusoid, a little tapered anteriorly and strongly recurved at the siphonal end; the aperture is elongate, subovate, widest above. The outer lip is thickened by a rib on the back, with a flaring callus above and coarsely lirated within. The inner lip margin has a narrow, smooth callous shelf and bears a keel at its end or at the point where the narrowed siphonal canal begins it bend. The posterior canal is enginoid and is represented by a large notch bordered by elevated edges, or lips, within the aperture at its upper end. Color dark brown or black.

The radula is a small, buccinoid ribbon, very long and narrow (length 2.4 mm, width 0.09 mm), with many rows of teeth (approximately 154). The central teeth are openly spaced, each tooth being a subrectangular plate with parallel sides, bearing three sharp primary cusps on the upper or cutting side, the middle one largest and with a small, secondary cusp on each side; lateral teeth are smaller, crowded and overlapping, each with a sharp, curved cusp at each end, the outer one the largest.

Measurements.—Length 22.0 mm, diameter 9.8 mm; holotype.

Holotype.—PILLSBURY Sta. P-522, Gulf of Panama, 7°37.8′N, 79°32.2′W, depth 276-272 m, 4 May 1967.

Remarks.—Distinguished from other species of *Engina* in the region by its fusoid shape. Only a single specimen taken at PILLSBURY Sta. P-522.

The species is named for Dr. James H. McLean of the Los Angeles County Museum, in recognition of his work on Californian and Panamic mollusks.

Genus Bayerius, new genus

Type-Species.—*Fusinus fragilissimus* Dall, 1908, here designated. Deep water off Ecuador.

Description.—The shell is fusiform, with whorls convex between deep sutures, the texture thin. The anterior canal is straight, narrow and pointed at the end, the siphonal canal not forming a fasciole. The aperture is semilunate, the pillar wall plain, the margin of the outer lip thin and without a sinusoidal inflexion. Sculpture consists of fine, alternating spiral threads, no axials. The radula is buccinoid, with a stout, rectangular central tooth bearing three cusps; the laterals are buccinoid, with three fanglike cusps, the outer a little larger.

Bayerius fragilissimus (Dall)
Figs. 19, 24, 25, 101

Fusinus fragilissimus Dall, 1908, Bull. Mus. Comp. Zool., *43*(6): 301, 302, pl. 12, fig. 6. (ALBATROSS Sta. 3398, off the coast of Ecuador in 1573 fathoms = 2877 meters. Type, USNM 123007.)

Remarks.—The shell is as described for the genus above. The holotype is a broken specimen 21.0 mm long. The sculpture of fine, alternating spiral threads is well shown and agrees perfectly with that of our specimens from the Gulf of Panama. By its short, narrow anterior canal with pointed siphonal end, *Bayerius fragilissimus* resembles in shape many other deepwater gastropods belonging to various families.

Material Examined.—PILLSBURY Sta. P-526, Gulf of Panama, 6°53'N, 79°27'W, depth 3193-3200 m, 5 May 1967, two specimens, UMML 30-7715.

Genus *Phymorhynchus* Dall, 1908

Type-Species.—*Pleurotomella castanea* Dall, by original designation.

→

FIGURES 28-44. Panamic Gastropoda: 28-30, *Terebra argosyia*, new species, holotype; 28, 29, ventral and dorsal aspects showing color pattern; 30, dorsal aspect of the same shell whitened to show sculpture; length 71.4 mm; ARGOSY Sta. 72.—31, a-b, 32, *Natica inexpectans*, new species; 31, a-b, holotype, length of shell 23.1 mm, diameter 22.6 mm; 32, operculum of same, outer surface, height 15 mm, width 10 mm; PILLSBURY Sta. P-513.—33, 34, *Truncaria filosa* (Adams & Reeve); 33, length 37.6 mm; 34, length 34.6 mm; PILLSBURY Sta. P-502.—35, *Oliva* sp.; side aspect of a specimen showing the uncolored labial band; length 33 mm.—36, 37, *Austrotrophon panamensis*, new species, holotype; length 32 mm, diameter 27 mm, PILLSBURY Sta. P-529.—38, 39, *Centrifuga swansoni* (Hertlein & Strong); 38, length 43.6 mm; 39, length 43.7 mm; ARGOSY Sta. 26.—40, 41, *Centrifuga centrifuga* (Hinds); ventral and posterior views, length 60.3 mm, PILLSBURY Sta. P-531.—42, *Engina macleani*, new species; length 22 mm, holotype, PILLSBURY Sta. P-522.—43-44, *Cantharus (Muricantharus) panamicus* (Hertlein & Strong); length 44.4 mm, diameter 29.7 mm, ARGOSY Sta. 26.

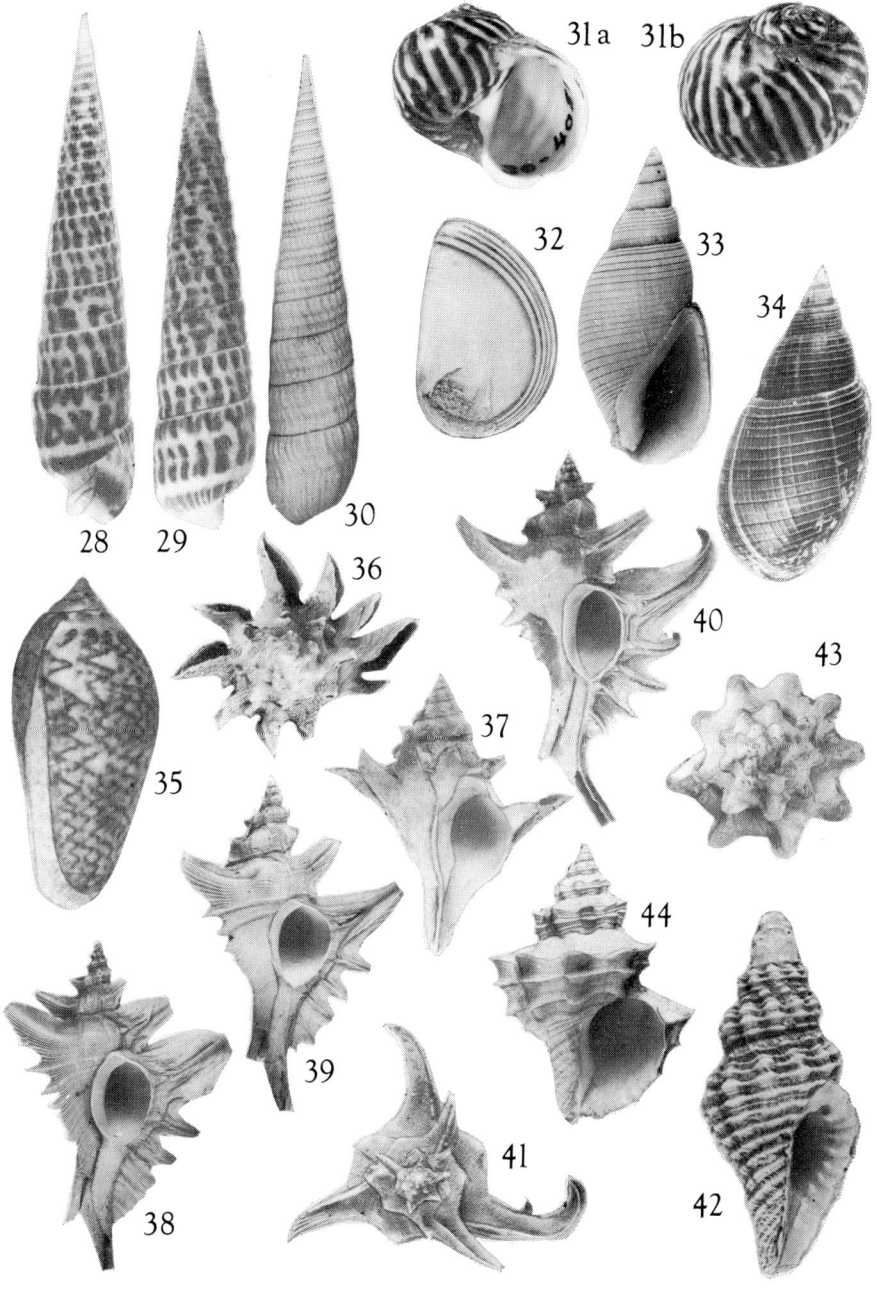

Description.—The shell is fusiform, with an elevated spire and rounded to slightly shouldered and spirally sculptured whorls. The anterior canal is narrow, straight, and terminally pointed. The margin of the outer lip is thin, straight, with the growth lines showing only a shallow insinuation at the suture. The parietal callous is spread narrow, thin. According to Dall, the animal is blind and has a distinct muzzle into which the proboscis is retracted. Radula unknown. Operculum lacking.

Remarks.—Dall classed *Phymorhynchus* as a subgenus of *Pleurotomella,* and hence a turrid. Certain Japanese species referred to this genus have, on radular characters, been shown to be buccinids. The general contour of the shell, especially the shape of its narrow anterior canal, is similar to that of *Bayerius.*

Phymorhynchus speciosus, new species
Fig. 18

Description.—The shell is fusiform, with an elevated spire of about six rounded, weakly shouldered whorls; the anterior canal is of moderate length, straight, narrowing to a blunt point. The sculpture is as figured, formed by sharp, elevated spiral cords, those at the shoulder belt larger and more widely spaced, those on the body proper and across the base finer and more closely spaced, and much finer along the canal. On the subsutural zone, the spirals are much finer and the axial lines stronger and slightly bowed in the vicinity of the suture. The nuclear whorls are small, corroded, and generally terminate in a point. Aperture subelliptical, the margin of its outer lip thin, serrated by the ends of the spiral sculpture and, in the specimens examined, without a sinal notch. The anterior canal is relatively short, narrow, and sticklike in shape, tapered almost to a point, its siphonal canal narrow and not recurved. The pillar is smooth, coated with a narrow callous pad upward over the parietal wall. Texture thin, fragile; color of the shell is white under a pale yellow periostracum.

Measurements.—Length 39.7 mm, diameter 18.4 mm, holotype.

Holotype.—USNM No. 701159, PILLSBURY Sta. P-526, Gulf of Panama, 6°53′N, 79°27′W, depth 3193-3200 m, 5 May 1967.

Remarks.—Somewhat similar to *Phymorhynchus castaneus* (Dall), but wider with stronger sculpture and white color.

Genus *Cantharus* Roeding, 1798

Type-Species.—*Buccinum tranquebaricus* Gmelin, by subsequent designation: Suter, 1913. Recent, Indo-Pacific.

Description.—The shell is ovate, short, and stout, the body whorl large, widely shouldered, the space between the shoulder and suture wide, flat or sloping, generally without pronounced axial sculpture. The aperture is broadly ovate, its upper side straight, rectangular, the columellar side concave, smooth, with a pillar keel bordering the siphonal canal recurved at the end forming a beak and a siphonal fasciolar fold around a closed or slightly perforate axis. The sculpture consists of narrow axials between wider intervals, the whole overridden by spirals. The operculum is corneous, with an apical nucleus. The radula is an elongated ribbon with a tricuspid central tooth, the lateral teeth larger with large, fang-shaped cusps at each end and two smaller intermediate cusps.

Remarks.—Through the courtesy of Dr. Harald Rehder of the National Museum of Natural History, I am able to illustrate the radula of typical *Cantharus* (Fig. 95), which will facilitate closer examination of the cantharid species in the American fauna.

Subgenus **Muricantharus**, new subgenus

Type-Species.—*Pseudoneptunea panamica* Hertlein & Strong, here designated.

Description.—The shell is cantharid, wide at the shoulder and with a spire of medium height with seven or more whorls; the anterior canal is short, twisted, with a small, columellar plait along the curved edge of the siphonal canal, its recurved beak resulting in a very small, hardly noticeable fasciolar fold. Sculpture consists of narrow axial riblets (about nine) between wide intervals overridden by four lines of primary spiral cords (also widely spaced) forming sharp, triangular nodes (transverse) at their intersections, that at the shoulder largest. The whorls are sharply shouldered; the space between the shoulder and suture is wide, flat, or merely undulated by the subdued extension of the axials and marked mainly with minute, threadlike spirals. The aperture is widely ovate, the edge of its outer lip thin, smooth, or weakly lirated within. The periostracum is thin, deciduous and flaky, dull gray-brown, the shell beneath white except for the axial nodes which are stained a deep brown. The operculum is horny, with a terminal nucleus.

The radula is a buccinoid ribbon, fairly large, with widely spaced teeth, the central tooth high with straight, parallel sides, deeply concave base, and three sharp cusps on the cutting edge; the lateral teeth are much larger than the centrals, typically buccinoid in shape, the outer cusp is large, fanglike, the inner one smaller, wider, with serration along its inner edge, and a smaller intermediate cusp. The ribbon extracted from the figured specimen (height of shell 45.0 mm, diameter 32.0 mm) has a length of about 13.0 mm, a width spread open of 0.41 mm, and about 146 rows of teeth (Fig. 97).

Cantharus (*Muricantharus*) *panamicus* (Hertlein & Strong)
Figs. 43, 44, 97

Pseudoneptunea panamica Hertlein & Strong, 1951, Zoologica, *36*(2): 81, pl. 2, figs. 6, 10. (Holotype, on Hannibal Bank, Panama, in 35-40 fathoms = 65-82 meters; other specimens from Gulf of California.)

Remarks.—With the characters as described for the genus. A distinctive species, easily recognized by the shape and sculpture of the shell. The radula extracted from the figured specimen has a length of about 13.0 mm, and a width of 0.41 mm when spread open. It shows about 146 rows of teeth and is typically buccinoid, but with sharp serrations along the inner edges of the lateral teeth.

Material Examined.—PILLSBURY Sta. P-529, Bay of Panama S of Isla de San José, 8°00.7'N, 79°11.8'W, depth 84 m, 6 May 1967, UMML 30-5493, 3 specimens.—PILLSBURY Sta. P-530, Bay of Panama W of Isla de San José, 8°15.3'N, 79°10.6'W, depth 66-71 m, 6 May 1967, UMML 30-4187, 2 specimens.—PILLSBURY Sta. P-550, Bay of Panama W of Punta Escondido, 8°00.7'N, 78°40.3'W, depth 64-60 m, 7 May 1967, UMML 5148, 1 specimen.—ARGOSY Sta. 26, NW of Buenaventura, Colombia, 4°00'N, 77°30'W, depth 91 m, 17 September 1961, 1 specimen (illustrated).

Genus *Truncaria* Adams & Reeve, 1850

Type-Species.—*Buccinum filosum* Adams & Reeve, 1848, by monotypy. Erroneously reported from the China Sea; most likely Panama.

Remarks.—This is a curious genus and readily recognizable by the shape of the shell, its smoothish, spirally lined sculpture, and especially by the oblique truncation of its columellar end; these features impart to the shell a remarkable resemblance to some land snails, such as *Euglandina*. Only a single species is so far known, *Truncaria filosa*, originally described as from China, but now believed to have come from Panama along with other West American species described in the Samarang Report as from the Eastern Seas. (For discussion, see Pilsbry & Olsson, 1935: 119, 120, *Tumbeziconcha*.) The radula is unknown, but the position of this genus amongst the Buccinidae seems well founded. There are fragments of a shell suggestive of *Truncaria* in the Caribbean collection of the Rosenstiel School of Marine and Atmospheric Sciences, which, when additional specimens are obtained, may prove to be a second species of this unique genus.

Truncaria filosa (Adams & Reeve)
Figs. 33, 34

Buccinum filosum Adams & Reeve, 1848, Zoology of the Voyage of the H.M.S. Samarang . . . Mollusks: 33, pl. 11, fig. 18. (Hab. China Sea.)

Truncaria filosa (Adams & Reeve), H. and A. Adams, 1853, Genera of Recent Mollusca, vol. 1: 111, 112, pl. 12, fig. 2.

Cominella brunneocincta Dall, 1896, Proc. U. S. Nat. Mus., 18(1034): 11, 12. (U. S. Fish Commission station 3390, in 56 fathoms, sand; temperature 62.6°F; in the Gulf of Panama. USNM 123009.)

Truncaria brunneocincta (Dall), Dall, 1908, Bull. Mus. Comp. Zool., 43(8): 304, 305, pl. 2, fig. 6.

Remarks.—The type-specimen of *Buccinum filosum* is in the collection of the British Museum (Nat. Hist.) in South Kensington and measures 29.0 mm in length and 11.9 mm in diameter. Close comparison of the type-specimen of *T. filosa* with my photographs of *Truncaria brunneocincta* from Panama show no specific differences, and it can be safely assumed that it originally came from Panama and not from the China Seas as originally reported. The three PILLSBURY specimens, dead and more or less encrusted, were trawled off Punta Mala, Los Santos Peninsula, Panama.

Material Examined.—PILLSBURY Sta. P-502, Gulf of Panama off Punta Mala, 7°40'N, 79°50.5'W, depth 79-77 m, 2 May 1967, UMML 30-7708, 3 specimens.

Genus Strombinophos Pilsbry & Olsson, 1941

Type-Species.—*S. loripanus* Pilsbry & Olsson, by original designation. Pliocene of Ecuador.

Description.—The shell is fusiform in shape and sculpture. The aperture is columbelloid, the anterior canal straight, without a pronounced siphonal fasciolar fold and keel. The radula is buccinoid, the central teeth semicircular in shape, multicuspid, the laterals buccinoid with the two ends produced as narrow, curved and fanglike cusps.

Remarks.—The genus *Strombinophos* is easily differentiated from other *Phos*-like genera by its characteristic shape, and this differentiation is now confirmed by its radular characteristics. *Trajana* has a similar radula.

Strombinophos fusoides (C. B. Adams)
Fig. 100

Remarks.—*Strombinophos fusoides* is a common, inshore species, especially near the entrance to the Panama Canal, and it was also taken by the PILLSBURY at ten stations in the offshore waters of the Gulf of Panama.

Material Examined.—Numerous specimens from the following PILLSBURY stations:

	Latitude	Longitude	Depth (m)	Date, 1967	Specimens	UMML No.
P-485	8°26.2'N	79°43.2'W	15-11	1 May	3	30-3786
P-486	8°20'N	79°49.7'W	20	1 May	3	30-3871
P-487	8°18.1'N	80°00.5'W	18	1 May	1	30-3795
P-488	8°13.1'N	80°09.6'W	17	1–2 May	4	30-3806
P-494	7°49.3'N	80°00'W	46	2 May	4	30-3954
P-495	7°59.2'N	80°00.2'W	40-37	2 May	3	30-3976
P-496	8°09.7'N	80°00.9'W	33	2 May	9	30-4013
P-535	8°38.6'N	78°51.9'W	31	6 May	1	30-5083
P-541	8°30'N	78°49.2'W	29	7 May	2	30-5094
P-546	8°19.2'N	78°35.8'W	27-31	7 May	1	30-5131

Family Muricidae

The family "Muricidae" is represented in the PILLSBURY and ARGOSY collections by numerous species, of which most are well-known forms and need no special mention in this paper. The family has been divided into several smaller categories or subfamilies, whose limits are not always clearly defined and of which the Muricinae (the true muricids) and the Ocenebrinae (the false muricids) include the majority of named genera. In these two major groups, differentiation is best observed in the characters of the radula, which show differences so great that there is a question as to how closely these two groups are related. In the Muricinae, the radular ribbon is of simple construction, fairly large and wide, and it resembles that of some other rachiglossate families, particularly the Olividae. In the Ocenebrinae, the ribbon is small, narrow, extremely long (threadlike), with a very large number of transverse rows (200-400) of teeth in which those of the central series are crowded together. The proper illustration of the ocenebrid radula, as well as its final interpretation, is a difficult matter demanding skill and experience.

Subfamily Ocenebrinae

Shell of varied form, high spired, with or without varices at regular intervals. The best criterion is furnished by the radula which is small, threadlike, and often very long, with small teeth in many closely adjacent and overlapping rows; the central tooth is usually in the form of a complex, rectangular, cusped plate and the lateral teeth are much smaller, claw- or thorn-shaped.

Genus *Centrifuga* Grant & Gale, 1931

Type-Species.—*Murex centrifuga* Hinds, 1844, by original designation. Panamic-Pacific province.

Description.—The shell is muricid in form, with elevated spire of angled or shouldered whorls bearing three narrow or winged varices extended into recurved spines at the shoulder, their forward side seamed or fluted. The aperture is ovate, closed by a rimmed peristome. The anterior canal is fusoid, bent sharply in its anterior section and roofed over along most of its length. The radula is small, threadlike, and very long, with many (200-300) transverse rows of teeth; the serially overlapping central teeth are complex in structure and design; the lateral teeth are much smaller and claw- or thorn-shaped.

Remarks.—*Centrifuga* is herein treated as a distinct genus, represented in the collections by two species from the Gulf of Panama. Relationship with *Pteropurpura* Jousseaume, 1880, has been suggested by some authors.

In 1960, Emerson contributed a special paper on some disputed eastern Pacific muricid gastropods in which the generic placement and identity of *Murex centrifuga* Hinds, the type of the genus *Centrifuga*, is critically discussed with illustrations of various specimens. The paper adds much new information and represents a praiseworthy study of a subject traditionally based on the variables in shell characters. The specimen figured by Hinds in the report on the mollusks of the Voyage of the Sulphur is missing from the type collection at the British Museum, as indicated by Myra Keen. Emerson indicated that a specimen (No. 79905) from the John Calvert collection at the American Museum of Natural History, which bears the label as being from the west coast of Veragua, the locality given by Hinds, appears to be the missing specimen. This specimen has been designated as a lectotype. The original figures of *Murex centrifuga* are reproduced at an enlargement of 2× and serve as a convenient means of comparison for the general worker who does not have access to Hinds's paper on the SULPHUR mollusks. Emerson also discussed the species described by Hertlein & Strong as *Pterynotus* (*Pteropurpura*) *swansoni* which, in spite of its more varied form and greater alation of its varical spines, was judged to be a junior synonym of *M. centrifuga*. Also of interest is a figure of the specimen (USNM 123019) mentioned but not figured by Dall (1908). In my opinion that specimen represents *C. swansoni*. It is clearly evident that there are two distinct species of *Centrifuga* in Panamic waters, distinguishable on both shell and radular characters.

Centrifuga centrifuga (Hinds)
Figs. 40, 41, 102

Murex centrifuga Hinds, 1884, Zool. Voy. Sulphur, Mollusca: 8, pl. 3, figs. 7, 8 (west coast of Veragua in 52 fathoms).—Sowerby, G. B. (2nd), 1879, Thes. Conch., 4: *Murex*, p. 25, no. 110, pl. 390, fig. 101.
Murex centrifuga Hinds, Emerson, 1960, Am. Mus. Novitates, No. 2009: 5,

6, fig. 1 (two enlarged figures after Hinds); fig. 2 (lectotype, AMNH No. 799-3).
Not *Murex* (*Alipurpura*) *centrifuga* Hinds, Dall, 1908, Bull. Mus. Comp. Zool., 43(6): 313 (= *Centrifuga swansoni* Hertlein & Strong).

Description.—The shell is muricid with an ovate aperture whose margin is closed and rimmed by a raised peristome; the anterior canal is straight, except for its recurved tip, which likewise is closed or roofed over completely. The varices, three in number, are narrowly flattened or expanded along the sides, toothed by the ends of obscure spiral cords, and produced at the shoulder into large, long, narrow spines recurved at their ends. The general color is grayish yellow; there is a light straw-colored periostracum. The operculum is chitinous, concentric, with a subapical nucleus. The radula is ocenebrid, very long, narrow and threadlike, with a large number of rows of teeth; the central teeth are large, crowded, overlapping, and cusped, and the laterals are small and claw-shaped (Fig. 102). A radular ribbon measured as follows: length 9.36 mm, width 156 mm; number of rows of teeth about 288.

Material Examined.—PILLSBURY Sta. P-531, Gulf of Panama, 8°25.5′N, 79°10.7′W, depth 57-64 m, 6 May 1967; one specimen, length 60.3 mm, diameter 40.0 mm.—Gulf of Panama, taken by shrimp trawlers, one specimen, length 65.7 mm, diameter 48.4 mm. (Olsson collection.)

Centrifuga swansoni (Hertlein & Strong)
Figs. 38, 39, 103

Pterynotus (*Pteropurpura*) *swansoni* Hertlein & Strong, 1951, Zoologica, 36(2): 85, pl. 2, figs. 8, 12. (Gulf of California.)

Description.—The shell is muricid, small or medium in size (average height 43 mm), with three winged varices produced into flattened, recurved spines at the shoulder of the body whorl and ascending the later whorls of the

→

FIGURES 45-56. Panamic Gastropoda: 45-46, *Calliostoma pillsburyae*, new species, holotype; two views of the same specimen, stained and whitened; height 17.3 mm, diameter 19.5 mm, PILLSBURY Sta. P-530, Gulf of Panama.—47-49, *Calliostoma veleroae* McLean; 47, 48, two views of the same specimen stained and whitened; 49, same specimen moistened, showing color pattern; height 18.6 mm, diameter 19.2 mm, UMML 30-5183, PILLSBURY Sta. P-555, Gulf of Panama.—50-51, *Calliostoma sanjaimense* McLean; two views of the same specimen, stained and whitened; height 24.2 mm, diameter 23.7 mm, UMML 30-7894, PILLSBURY Sta. P-501.—52, 53, *Calliostoma joanneae*, new species, holotype; two views of the same specimen, stained and whitened; height 8.7 mm, diameter 7.7 mm, PILLSBURY Sta. P-500.—54-56, *Calliostoma decipiens*, new species, holotype; 54, 55, two views of the same specimen, stained and whitened; 56, natural, to show color pattern; height 13.2 mm, diameter 13.5 mm; PILLSBURY Sta. P-556.

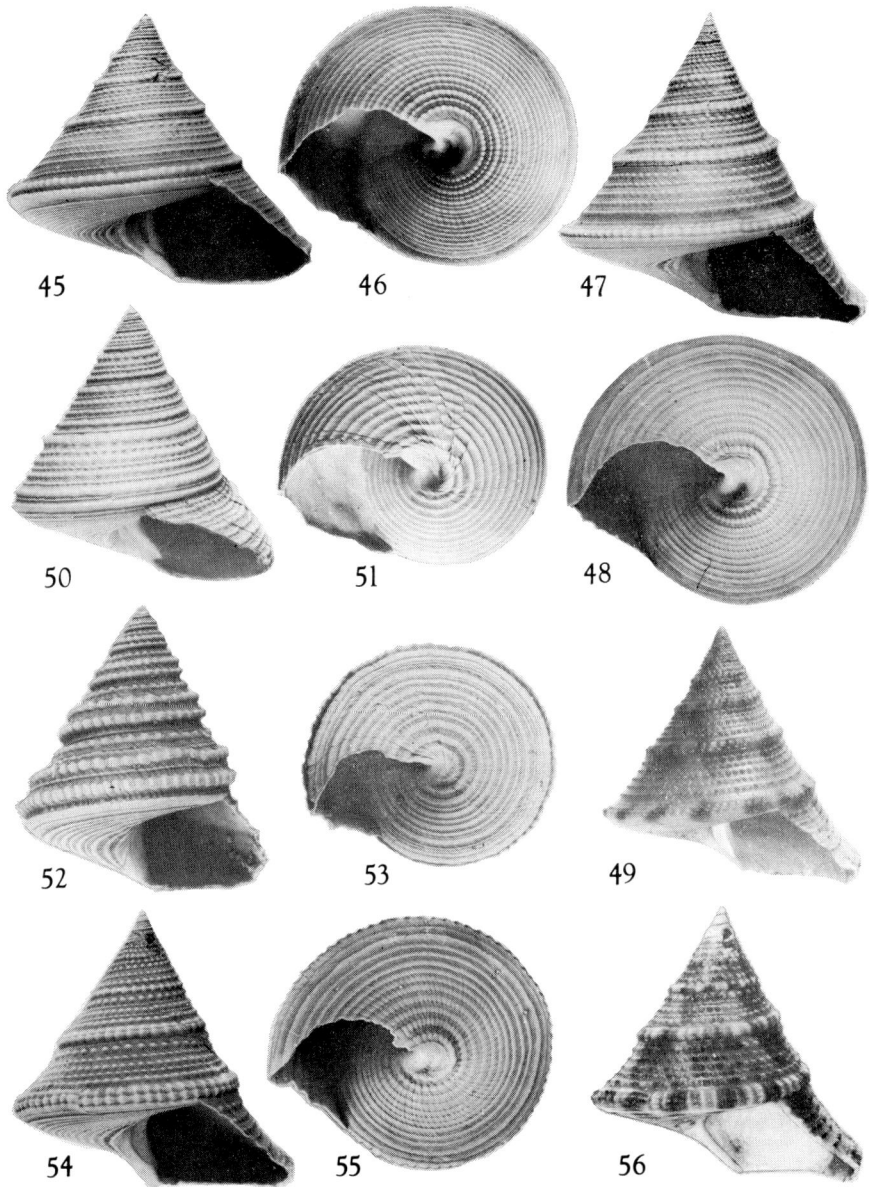

spire. On the body whorl, the varices extend from the angle of the shoulder as narrower blades, their edges coarsely toothed by the ends of a set of primary, ridgelike spiral cords. The spire is elevated, composed of seven or more whorls, the apical ones lacking in the available specimens. Intervarical surfaces bear a single knob in the center at the shoulder; the surfaces are smoothish or with fine, subdued spirals which increase in strength on the back of the winged varices. The aperture is ovate, with a raised rim. The anterior canal is sealed over, except in the juvenile form. Other characteristics are as described for the genus.

Comparisons.—Several specimens of this species from both the ARGOSY and PILLSBURY collections have been available. These show that shells of *C. swansoni* are generally smaller, with more variable characters, especially as to the extent that the varices are winged. The radular ribbon, extracted from several specimens, is smaller than that of *C. centrifuga,* its central tooth has more pronounced Ocenebrinid characters (Fig. 103), and its central cusp is placed in the middle of the plate. The radula bears a general resemblance to that of *Ocenebra erinaceus* Linné, as figured by Troschel (Gebiss der Schnecken, vol. 2; pl. 11, figs. 11, 12).

Measurements.—Sizes of three specimens are as follows: length 43.9 mm, diameter 29.6 mm; length 44.0 mm, diameter 29.0 mm; length 42.0 mm, diameter 29.0 mm.

Measurements of two radulae are as follows: length of ribbon 9.8 mm, width 0.13 mm, rows of teeth about 300; length of ribbon 7.3 mm, width 0.12 mm, rows of teeth about 244.

Material Examined.—PILLSBURY Sta. P-502, Gulf of Panama NE of Cabo Mala, 7°40′N, 79°50.5′W, depth 79-77 m, 2 May 1967, UMML 30-4066, 9 specimens.—PILLSBURY Sta. P-529, Gulf of Panama S of Isla de San José, 8°00.7′N, 79°11.8′W, depth 84 m, 6 May 1967, UMML 30-4168, 11 specimens.—ARGOSY Sta. 26, NW of Buenaventura, Colombia, 4°00′N, 77°30′W, depth 91 m, 17 September 1961, 3 specimens.

Genus *Austrotrophon* Dall, 1902

Type-Species.—*Trophon cerrosensis* Dall, 1891, by subsequent designation: Grant & Gale, 1931. Cerros Id., Lower California to the Gulf of California.

Austrotrophon panamensis, new species
Figs. 36, 37

Description.—The shell is of medium size with an elevated spire of seven or eight shouldered whorls and a short, straight, or slightly curved, anterior canal. The shape of the shell is shown in the figure. The sculpture is formed by deeply fluted varices, about nine on the body whorl, which are most strongly developed as a crown of large, broad spines along the shoulder of the body whorl; the rest of the surface is smooth except for axial lines of growth. The anterior canal is quite short, straight, narrower below, with an indistinct umbilical chink, and slightly recurved.

Measurements.—Length 32.0 mm, diameter 27.0 mm; holotype.

Holotype.—USNM No. 701162. PILLSBURY Sta. P-529, Gulf of Panama S of Isla de San José, 8°00.7′N, 79°11.8′W, depth 84 m, 6 May 1967.

Remarks.—Related to the more northerly *A. cerrosensis* (Dall), it differs by its stubbier form, the larger fluted varices restricted to the shoulder, and its plain surface devoid of any spirals.

Superorder MESOGASTROPODA
Order TAENIOGLOSSA
Superfamily Naticacea
Family Naticidae

Genus *Natica* Scopoli, 1777

Type-Species.—*Nerita vitellus* Linné (="*Natica rufa* Born" of authors), by subsequent designation: Harris, 1897. Recent, western Pacific.

Description.—The shell has a large, convex body whorl and generally a low spire of rapidly diminishing whorls upward. The general surface is smooth, but often with incised tangential sulci at the suture. The aperture is large, oblique, semilunate in shape, its pillar side straight. The operculum is calcareous, its external face with semicircular ribs.

Natica inexpectans, new species
Figs. 31, 32

Description.—The shell is small, fairly solid, with a large, convex body whorl and a low spire of about four whorls. The suture is fine and sharp, not bordered by tangential sulci, but the growth lines along the suture are a little coarser there than over the rest of the surface. The umbilicus is open and deep, and the funicular rib on the pillar side is very small or wholly lacking. The surface is smooth and glossy beneath a thin, yellowish periostracum. The aperture is broadly semilunate, the margin of the outer lip inclined backward, and the pillar side straight and smooth. The color is white with narrow brown bands parallel to the margin of the lip and extending unbroken from the edge of the umbilicus upward to the suture; in some specimens the bands have one or more angular flexions in the vicinity of the shoulder.

The calcareous operculum is almost flat, semicircular in outline and

Tropical American Mollusks

FIGURES 57-70. Panamic Gastropoda: 57-60, *Calliostoma nepheloide* Dall; the same specimen in natural form to show general pattern of coloration, and whitened to show details of sculpture; height 18.7 mm, diameter 20 mm, UMML 30-4033, PILLSBURY Sta. P-500.—61, *Calliostoma iridium* Dall; an immature specimen, height 6.6 mm, PILLSBURY Sta. P-512.—62, 63, *Calliostoma insignis*, new species, holotype; two views of same specimen, whitened; height 15.2 mm,

tightly fitting the aperture of the shell, with a small spiral nuclear coil, its outer surface weakly concave, white and highly polished except for a discolored rugose area in the region of the nucleus; the outer edge is rimmed by four smooth, narrow, raised marginal ribs, the outer one forming the edge.

Measurements.—Length 23.1 mm, diameter 22.6 mm (holotype); length 14.2 mm, diameter 15.8 mm; length 16.5 mm, diameter 17.3 mm; length 16.8 mm, diameter 17.0 mm (all paratypes).

Holotype.—USNM No. 701161, PILLSBURY Sta. P-513.

Material Examined.—One live-collected and four dead specimens from the following two stations of R/V PILLSBURY: P-513, Gulf of Panama ENE of Cabo Mala, 7°39.5′N, 79°40.7′W, depth 117 m, 4 May 1967, one specimen alive when collected (holotype). P-521, Gulf of Panama NE of Cabo Mala, 7°48.3′N, 79°35.1′W, depth 119 m, 4 May 1967, four dead but quite fresh specimens (paratypes, UMML 30-7709).

Family Cypraeidae

The cowries taken by the PILLSBURY and ARGOSY during their cruises in Panamic waters southward to Isla la Plata comprise four species, *C. cervinetta, arabicula, robertsi* and *albuginosa*. *C. arabicula* is the commonest species, especially at, and in the general vicinity of, Isla la Plata. *C. albuginosa*, taken only by the ARGOSY, is of normal size and needs no further comments. Most specimens of the other species are abnormally small, even when mature, and seem to show that the environmental conditions in deeper, offshore waters were less favorable than at shallower, inshore stations. It was noticed that the shape and coloration of the juvenile stages of these cowries are distinctive and provide means by which young shells can be specifically recognized. Also included herein are descriptions and figures of specimens of the rare *Cypraea aequinoctialis* Schilder, collected at Manta, Ecuador.

Cypraea (Zonaria) arabicula (Lamarck)

Cypraea arabicula Lamarck, 1810, Ann. Mus. Hist. Nat. Paris, *16:* 100.—
  Sowerby, G. B., 1870. Thes. Conch., *4: Cypraea*, p. 16, pl. 298, figs. 38, 39 (St. Elena. West Colombia).

←

diameter 15.3 mm, PILLSBURY Sta. P-502.—64, 65, *Cypraea robertsi* Hidalgo; two figures of the same adult specimen; height 18.0 mm, diameter 12.4 mm, PILLSBURY Sta. P-543.—66-68, *Cypraea aequinoctialis* Schilder; different aspects of the same specimen; length 45 mm, Manta Ecuador, Olsson collection.—69, 70, *Cypraea robertsi* Hidalgo; an immature specimen showing shape and speckled color pattern; length 23 mm, diameter 16 mm, ARGOSY Sta. 59.

Zonaria (Zonaria) arabicula (Lamarck), Cate, 1969, Veliger, *12*(1): 115, pl. 13, fig. 15. (Range: Gulf of California to northern Peru and the Galapagos.)

Description.—The shell is small to medium in size, solid, and when mature ovately elliptical in shape so as to appear almost rectangular, the dorsal surface high, convexly domed, the base flattened, the two sides (those of the body and the lip) of almost equal width, the lateral sides calloused and angled. The aperture is narrow and nearly straight, except for a short slant to the left forming the siphonal canal. The apertural teeth on the inner edge of the outer lip are small, even, numerous (about 23), and those on the parietal edge are longer, riblike, extending inward, and less numerous (about 18). The fossula is large and deep, its center usually smooth, with coarse teeth on the inner and outer sides. The color is a rich brown due to a thick sprinkling of small, brown dots, generally confluent to form small figures on a blue-gray base. A narrow, nearly straight, mantle line is usually evident a little displaced towards the lip side, and often there are faint traces of three or more cross bands. The base is white or pink, plain or with brown spots spreading inward from the sides. The lateral margins are coated with a deposit of callus, usually pinkish and bearing a row of large, brown spots along the carinate edge and upward to the edge of the calloused coat.

Measurements.—A series of mature shells from Ecuador and Peru have the following measurements: length 30.3 mm, height 15.8 mm, diameter 20.6 mm (Esmeraldas, Ecuador); length 27.4 mm, height 14.5 mm, diameter 18.7 mm (Isla la Plata, Ecuador); length 30.9 mm, height 16.0 mm, diameter 20.8 mm (Pena Malo, Peru); length 33.0 mm, height 16.8 mm, diameter 20.9 mm (Mancora, Peru).

Remarks.—*Cypraea arabicula* is a fairly common intertidal species sharing the same ecological environment as *C. robertsi*. Juvenile specimens of *C. arabicula* have rounded sides, and the color is usually bluish gray overlain by a pattern of close-set revolving lines of small brown spots which extend down onto the base of the body-whorl. Such juvenile specimens resemble young specimens of *C. robertsi*, but can be separated by the shape of the aperture and numerous small labial teeth.

Cypraea (Zonaria) robertsi Hidalgo
Figs. 64, 65, 69, 70

Cypraea punctulata Gray, 1824, Zool. Journ., *1:* 387 (not of Gmelin, 1791).
  —Sowerby, 1832, Conch. Illust.: *Cypraea*, pl. 4, fig. 20.
Cypraea robertsi Hidalgo, 1906, Mem. Acad. Cienc. Madrid, *25:* 176.—Cate 1969, Veliger, *12*(1): 114, fig. 13.

Description.—The shell is medium sized, solid, ovate, the apical depression small (in the juvenile showing the tip of the spire), dorsum roundly convex and sloping into the sides. The posterior terminals of the aperture are barely produced. Lateral sides have a band of callus, heaviest below, thinning as it spreads upward about half way on the sides. The aperture is narrow, sinuous, curving toward the left above and expanding a little over the fossula. The fossula is shallow and shows mainly as a groove cutting across the bordering teeth, the ends of which barely meet. The color of adults is variable, from a bluish gray to brown, with or without darker cross bands and a sprinkling of small, confluent dots, the terminal ends with a dark brown spot. The lateral bands of callus may be quite light or dark brown with darker brown spots over them. The base is white to pinkish with spots on the rounded sides. The pattern of juveniles is bluish gray streaked with a pattern of fine, transverse lines which extend also down over the body section of the base.

Remarks.—A common species, mainly in intertidal waters where it is usually associated with *C. arabicula*. It is distinguished from that species by its more rounded sides, apertural characters, and in the juvenile stages by a finely speckled pattern, as shown by Figures 69 and 70.

Cypraea (Zonaria) aequinoctialis Schilder
Figs. 66-68

Zonaria (Zonaria) annettae aequinoctialis Schilder, 1933, Zool. Anz., *101:* 193, fig. 3.—Cate, 1969, Veliger, *12*(1): 113, 114, pl. 12, fig. 12 (Cabo San Lorenzo, Ecuador).
Cypraea aequinoctialis (Schilder), Burgess, 1970, The Living Cowries: 348, No. 177, pl. 42, fig. B (off Taboga Island, Bay of Panama).

Description.—The shell is medium to large in size, subsolid, ovate to pyriform, the dorsum high, convex, and evenly rounded into the sides, the apical depression small. The aperture is sigmoidal, fairly wide, no fossula; teeth along the inner edge of the outer lip fairly large and of uniform size (about 18), those on the parietal edge longer, extending into the interior (about 16). Immature specimens have three wide bands across the dorsum, the middle one of which is wide and double, on a blue-gray base. Even in the banded stage, a wide band of callus spotted with large, brown dots extends up along both sides. The adult pattern over the dorsum consists of a close sprinkling of confluent brown spots, the cross bands showing faintly. The base is plain white or yellowish brown flushed with crimson.

Measurements.—Length 45.0 mm, height 22.0 mm, diameter 27.2 mm, with mature pattern (Manta, Ecuador); length 42.0 mm, height 21.2 mm, diameter 27.0 mm, with banded pattern over dorsum (Manta, Ecuador);

length 26.0 mm, height 13.0 mm, diameter 17.0 mm, mature pattern (Mancora, Peru).

Remarks.—A rare species not represented in the PILLSBURY or ARGOSY collections, but introduced here for comparison with *C. nigropunctata* Gray (Peru and Galapagos), with which it has sometimes been confused. It differs from *C. annettae* Dall in shape and juvenile color pattern.

Superorder ARCHAEOGASTROPODA
Order RHIPIDOGLOSSA
Superfamily Trochacea
Family Trochidae
Subfamily Trochinae

Genus *Panocochlea* Dall, 1908

Type-Species.—*Clanculus* (*Panocochlea*) *rubidus* Dall, by original designation.

Description.—The shell is small, rounded, with a low turbinate spire of about four whorls separated by sharp sutures, its surface sculptured with beaded spirals. The base is flattened, smooth, nearly covered with a glossy, pearly callus covering the umbilicus. The aperture is oblique, its outer lip sharp above, thickening along its basal side and merging into the parietal callus. The umbilical region is slightly pitted or flat, with a short, curved liration terminating at the aperture in a small tooth. The operculum is chitinous. The radula is rhipidoglossate, with compact, bushy marginal teeth; the middle section of the ribbon is rather wide, with a central tooth that is relatively large, with a round summit lacking an evident cusp, and with five much smaller, cusped lateral teeth on each side.

Remarks.—It deserves notice that the original material available to Dall was inadequate, consisting of a small, immature shell and an eroded fragment of a larger specimen. Nevertheless, Dall's praiseworthy diagnosis is essentially complete and testifies eloquently to the knowledge and skillful interpretation of its famous author.

Panocochlea rubida Dall
Figs. 78, 79

Clanculus (*Panocochlea*) *rubidus* Dall, 1908, Bull. Mus. Comp. Zool., *43*(6): 346, 347, pl. 8, figs. 3, 4. (ALBATROSS Sta. 3355, Gulf of Panama, in 182 fathoms = 333 meters, mud; USNM 122953.)

Description.—The shell has characters as described for the genus. Sculpture on the slightly domed dorsal surface consists of five beaded spirals, of which, the one bordering the suture is largest. The ventral surface is essentially smooth, with a spreading pearly callus, thickest over the parietal zone; the riblike lira bordering the lower side of a slight umbilical depression are short, curved, terminating at the aperture in a very small tooth. The color is a creamy white with only a slight tinge of red. The operculum was not observed, as the soft parts in two specimens were deeply retracted within the shell and it was necessary to use the KOH maceration process for extraction. The radular ribbon is quite small and demands the use of high power for detailed visual examination. A ribbon extracted from one specimen measures: length about 3.1 mm, width fully spread open 0.4 mm; it contained about 64 rows of teeth. The general aspect of the ribbon, with its bushy marginals and wide middle area, shows some resemblance to the radula of *Calliostoma* and of *Clanculus* as figured by Troschel (but with marked differences), and supports the assignment of the genus to the family Trochidae. The most conspicuous feature of the ribbon is the large size of the central tooth in relation to the small laterals, and the peculiar striped pattern of the marginals.

Measurements.—Height 9.3 mm, greater diameter 10.8 mm.

Material Examined.—PILLSBURY Sta. P-522, Gulf of Panama ENE of Cabo Mala, 7°37.8′N, 79°32.2′W, depth 276-272 m, 4 May 1967, UMML 30-7710, 3 specimens.

Subfamily Calliostomatinae

Genus *Calliostoma* Swainson, 1840

Type-Species.—*Trochus conulus* Linné, by subsequent designation: Herrmannsen, 1846. Recent, Mediterranean.

Description.—The shell is trochoid, with a conic spire and generally a flattened base, umbilicate or imperforate; the inner layer is thick and nacreous, the outer thin and sculptured. The aperture is inclined, rounded to oblique, the outer lip not thickened, either smooth or lirate within. The base of the pillar is thickened and forms a short tooth. Surface sculpture is mainly spiral, consisting of smooth or beaded cords. The operculum is corneous, circular, and multispiral. The radula is rhipidoglossate, usually quite large, with a bushy complex of marginal teeth in two sets, the lateral and central teeth together forming a wide, inner zone, the single central tooth generally bordered by five laterals (four to seven), prominently cusped, the edges of the cusps neatly serrated or fimbriated.

Remarks.—Authors have generally accepted Herrmannsen's type-designation of *Trochus conulus* Linné as valid, since the *Trochus conula* Mart. of Swainson's list represents the Linnean species, as pointed out by Hanley.

On the other hand, to accept as type (monotypic) *Calliostoma zizyphinus* Linné, mentioned singly on an earlier page by Swainson, as accepted by Clench & Turner (1960), seems to be an unnecessary and forced interpretation (see Woodring, 1928).

Clench & Turner (1960), in their review of the Calliostomas of the western Atlantic, defined the genus broadly to include both umbilicate and nonumbilicate species, as some forms that are openly umbilicated in the juvenile stage become imperforate in the adult. The radula is quite similar in all the species examined so far. The membranous jaws are said to be diagnostic of major groups or subgenera and can generally be obtained unharmed, attached to the radular ribbon, if the alkaline extraction has not been unnecessarily prolonged.

The Calliostomas of the PILLSBURY cruise comprise six species, of which four are described herein as new. The lovely *C. iridium* Dall is represented by a single immature specimen.

Calliostoma nepheloide Dall
Figs. 57-60

Calliostoma nepheloide Dall, 1913, Proc. U. S. Nat. Mus., *45:* 592 (Panama Bay in 47 fathoms).—McLean, 1970, Veliger, *12*(4): 422, pl. 62, figs. 2 to 5.

Not *Calliostoma nepheloide* Dall, 1925, Proc. U. S. Nat. Mus., *66*(17): 9, pl. 24, figs. 2, 3. (The figure represents another species introduced by mistake and not a reconstructed specimen as stated by McLean.)

Description.—The shell is of moderate size, conic (as shown in Figures 1-4), imperforate, and thin. The periphery of the final whorl is sharply carinate, bearing a single small, primary, beaded spiral and bordered just below and a little inside by a smaller plain or unbeaded spiral cord so that the peripheral angle is falsely bicarinate. The surface of the final whorl above the peripheral spiral is flat or slightly convex (as in the holotype) and sculptured with about 12 finely beaded spirals, which may be of uniform size and fairly widely spaced or more irregular. On the whorl above the last, the spirals are divided into two sizes: a series of six primaries separated by much smaller secondary or tertiary threads in their intervals. On the base, the spirals are smaller, more numerous (about 15), the outer

→

FIGURES 71-80. Panamic Gastropoda: 71-74, *Pectinodonta gilbertvossi*, new species; 71, 72, length 24.9 mm, paratype; 73, length 26.4 mm, holotype; 74, radula; PILLSBURY Sta. P-526.—75, 76, *Turcica panamensis*, new species, holotype; length 31.0 mm, diameter 29.0 mm; PILLSBURY Sta. P-529.—77, *Cirsotrema togata* (Hertlein & Strong); length 22.4 mm; PILLSBURY Sta. P-543.—78, 79, *Panocochlea rubida* Dall; length or height 9.3 mm, diameter 10.8 mm; PILLSBURY Sta. P-522.—80, *Pectinodonta nassa* (Dall), holotype; length 8.5 mm; USNM No. 123053.

ones narrow and plain, the inner more distinctly beaded. General color is an olivaceous brown modified by a pattern of large, brown flammules arranged axially across the surface (about eight on the last whorl) in alternating fashion, and with a row of smaller brown dots (about 16) along the peripheral zone. The nearly flat base is lighter in color, without flammules, and with a sprinkling of small brown dots on the primary spirals.

Material Examined.—Specimens from six PILLSBURY stations, as follows:

	Latitude	Longitude	Depth (m)	Date, 1967	Specimens	UMML No.
P-500	7°59.7'N	79°49.7'W	53	2 May	17	30-4033
P-501	7°50.2'N	79°50.5'W	68	2 May	4	30-4057
P-502	7°40'N	79°50.5'W	79-77	2 May	2	30-4071
P-521	7°48.3'N	79°35.1'W	119	4 May	1	30-4125
P-530	8°15.3'N	79°10.6'W	66-71	6 May	1	30-4183
P-531	8°25.5'N	79°10.7'W	57-64	6 May	4	30-5017

Remarks.—This is the common offshore *Calliostoma* in the Gulf of Panama, and numerous live specimens were taken during the PILLSBURY cruise at several stations listed above. The species is easily recognized by its shape, unicarinate peripheral keel, and color pattern.

Calliostoma pillsburyae, new species
Figs. 45, 46

Description.—The shell is of moderate size, white without pattern, conic, with the diameter of the body whorl greater than the height of the spire. The angle of the periphery of the body whorl is bicarinate, bearing two beaded cords; the sutural coil follows along the lower one, hence only the upper one shows as a beaded cord bordering the suture on the spire. Profile of the surface between the beaded peripheral cord and the suture is flat, ornamented with small, beaded spirals (about six), the uppermost of which is the stronger on the body and penultimate whorls; on the prepenultimate whorl are three primaries and three secondaries in alternate arrangement (details of which are shown in the figures). The base is flat or slightly convex, imperforate, ornamented with circles of evenly spaced spiral cords (about 17), of which the outer ones are plain, the inner larger and beaded.

Measurements.—Height 17.3 mm, diameter 19.5 mm; holotype.

Holotype.—USNM No. 701212, PILLSBURY Sta. P-530.

Material Examined.—The holotype, from Sta. P-530, Gulf of Panama west of Isla de San José, Las Perlas, 8°25.5'N, 79°10.7'W, 57-64 m, 6 May 1967.

Calliostoma insignis, new species
Figs. 62, 63

Description.—The shell is of moderate size, conic, its shape as shown in Figures 62 and 63, solid, imperforate, and rather coarsely sculptured. The whorls number about six and are flat-sided; the apical ones have been lost on the type. The periphery of each whorl is unicarinate, bearing a single coarse noded cord forming a keel which, on the penultimate and earlier whorls, prominently overhangs the whorl in front; the suture follows a deep groove just below the peripheral cord. The surface between the periphery and the suture above is marked with smaller, alternating, beaded spirals. The base is flattened, sculptured with eight or more beaded primary cords with a secondary thread between some of them. Color is an olivaceous green mottled with brown flammules.

Measurements.—Height 15.2 mm, diameter 15.3 mm; holotype.

Holotype.—USNM No. 701213, PILLSBURY Sta. P-502.

Material Examined.—The holotype, from Sta. P-502 in the Gulf of Panama northeast of Cabo Mala, 7°40'N, 79°50.5'W, depth 79-77 m, 2 May 1967. —Also a paratype from Sta. P-556, Gulf of Panama, 7°50.5'N, 78°49'W, depth 59 m, 8 May 1967; UMML 30-5188.

Calliostoma joanneae, new species
Figs. 52, 53

Description.—The shell is small, high, conic, with about seven coarsely sculptured whorls separated by deeply excavated sutures. The sculpture consists of three noded primary spiral cords on each whorl; the first, or lowest, forms a blunt, but not prominent, peripheral keel on the body whorl and an overhand on the whorls of the spire; the second cord, somewhat smaller, encircles the middle of each whorl; and the third and smallest borders the suture above; the interval between each pair of primary cords carries a fine, spiral thread. The base is flattened, sloping inward toward the middle; it carries at its outer edge a prominent, smoothish cord along which the coiling proceeds, hence it is concealed in the sutures on the spire. A series of nine somewhat alternating cords extends from the pillar outward toward a wide, flat, marginal circle marked with three smaller raised threads.

Measurements.—Height 8.7 mm, diameter 7.7 mm; holotype.

Holotype.—USNM No. 701165, PILLSBURY Sta. P-500.

Material Examined.—The holotype, a single specimen without soft parts, from Sta. P-500, Gulf of Panama, 7°59.7'N, 79°49.7'W, 53 m, 2 May 1967.

Remarks.—This species is named for Mrs. Jo Anne Hendricks Romfh, formerly assistant in the Deep-Sea Biology Program at the Rosenstiel School of Marine and Atmospheric Sciences, in recognition of her willing help during my work at the school.

Calliostoma veleroae McLean
Figs. 47-49

Calliostoma veleroae McLean, 1970, Veliger, *12*(4): 423, 424, pl. 62, fig. 12. (Three miles south of Isla Ladrones, Panama, 54 fathoms = 99 meters, VELERO III sta. 943-39; holotype, LACM-AHF 1271.)

Description.—A distinctive species recognized by the deeply impressed profile of its body whorl, its color pattern, and sculpture. The peripheral angle is compressed, bicarinated, and bears two coarsely noded cords bordered just below by a plain cord emerging at the aperture from the suture; hence, the two coarsely noded peripheral cords are exposed as well on the spire whorls as on the body. Details of the shape and sculpture are shown by the figures.

Measurements.—Height 15.9 mm, diameter 17.0 mm, holotype (after McLean); height 13.3 mm, diameter 13.4 mm, UMML 30-5210.

Material Examined.—PILLSBURY Sta. P-555, Gulf of Panama, 7°50.7'N, 79°00.3'W, 68 m, 7 May 1967, UMML 30-5183, three specimens.—PILLSBURY Sta. P-561, Gulf of Panama, 7°40'N, 78°30'W, 81 m, 8 May 1967, UMML 30-5210, one specimen.

Calliostoma iridium Dall
Fig. 61

Calliostoma iridium Dall, 1896, Proc. U.S. Nat. Mus., *18*(1034): 7; 1902, Proc. U.S. Nat. Mus., *24*(1264): 552, pl. 39, fig. 3; 1908, Bull. Mus. Comp. Zool., *43*(6): 348, pl. 19, fig. 5.—McLean, 1970, Veliger, *12*(4): 421, 422, pl. 62, fig. 1 (with figure of holotype). (Gulf of Panama in 127 fathoms = 232 meters, USNM 122957, holotype; and sta. 3391, in 153 fathoms = 280 meters.)

Description.—The shell is small, delicate, high conic, with impressed, flat, or slightly convex sides. The periphery of the last whorl is bicarinated by neatly beaded spiral cords, while the shiny surface above may be smooth, except for a beaded spiral at the suture, or faintly marked with weak spirals as figured by Dall and McLean. The base has a spiral cord on its outer margin and weak ones in the middle, the surface between is smooth. The general color is pinkish, glossy, and with a waxy, pearly sheen marked by complete or interrupted flammules of brown beginning as two spots on the peripheral cords and axially inclined upward across the general surface; the base is unmarked.

Measurements.—Height 20.0 mm, max. diameter 18.0 mm, height of aperture 7.0 mm, holotype, USNM 122957.

Material Examined.—PILLSBURY Sta. P-512, Gulf of Panama E of Cabo Mala, 7°30.5′N, 79°41.5′W, 210-205.

Remarks.—This is a lovely, deep-water species with a brilliant pearly or iridescent surface luster, as noted by Dall. The species has been well figured by Dall and more recently by McLean. Our single specimen is small.

Calliostoma sanjaimense McLean
Figs. 50, 51

Calliostoma sanjaimense McLean, 1970, Veliger, *12*(4): 423, pl. 62, fig. 11. (San Jaime Bank, west of Cape San Lucas, Baja California, Mexico.)

Description.—The shell is of moderate size, stout, high, conic, flat-sided (shaped as shown in Figure 50), and imperforate. The periphery of the last whorl is bicarinate, bearing a plain primary spiral cord at the angle marking the outer circle of the base, and above is a beaded primary cord with a small or tertiary thread in between. The surface of the last whorl is flat and sculptured with six beaded primary spiral cords, alternating with beaded secondary ones between, of which the upper cord adjacent to the threadlike suture is conspicuously the larger. The base is nearly flat or with a slightly convex surface bearing a series of about ten narrow, smooth, circular cords of nearly uniform size and spacing. Color is yellowish brown, plain or weakly spotted or mottled along the peripheral zone.

Measurements.—Height 25.0 mm, diameter 23.1 mm, PILLSBURY Sta. P-501.

Material Examined.—PILLSBURY Sta. P-501, Gulf of Panama NNE of Cabo Mala, Los Santos Peninsula, 7°50.2′N, 79°50.5′W, 68 m, 2 May 1967, UMML 30-7894, one specimen.

Remarks.—This fine species was recently described from a few specimens collected off Baja California, and its discovery in the Gulf of Panama marks a noteworthy southward extension of range.

Calliostoma decipiens, new species
Figs. 54-56

Description.—The shell is of moderate size, with an elevated sharp spire; the general surface of each whorl is distinctly concave between the angled periphery and the suture. The periphery of the body whorl is acute and bears a single small, projecting, weakly beaded cord, with the coiling of the whorl following just below it so that the peripheral cord remains prominently exposed and elevated above the suture on the larger whorls of the spire. The surface between the peripheral keel and the suture is flatly depressed and neatly ornamented with about five small beaded spirals of equal size and spacing. The base is nearly flat, bearing several small, adjacent, circular cords, larger and beaded around the middle.

Measurements.—Height 13.2 mm, diameter 13.5 mm; holotype.

Holotype.—USNM No. 701223, PILLSBURY Sta. P-556.

Material Examined.—The holotype, from PILLSBURY Sta. P-556, Gulf of Panama, 7°50.5'N, 78°49'W, 59 m, 8 May 1967.

Remarks.—This species resembles *C. nepheloide* Dall, but differs by its lighter and thinner shell as well as in details of sculpture. It also resembles *C. veleroae* McLean in shape, but its peripheral keel is narrower, sharper, and bears a single, beaded cord. The color of the dry shell in the collection is chalky white, but when moistened a pattern is visible as illustrated in Figure 56.

Subfamily Margaritinae

Genus *Turcica* A. Adams, 1854

Type-Species.—*T. monilifera* A. Adams, by monotypy. Recent, Australia.

Remarks.—The relationship of *Turcica* to other trochoid genera is uncertain. Pilsbry (1889:414) classed it as a subgenus *Calliostoma*, from which it differs by its plain, indistinctly paucispiral operculum and in radular characters. Thiele (1929), followed by Wenz (1938), assigned it to the Margaritinae. The genus is readily recognizable by its massive, biplicated columellar pillar. The rhipidoglossate radula is fairly large, but cannot be fully characterized at this time. The marginal teeth are very numerous, bushy, and form two sets: an inner set with a tendency to incline inward, formed of long, narrow blades with serrated or pectinated distal ends; and an outer set of wider, expanded blades with rounded, spatulated ends. The rachidian tooth is relatively high, with a sharp cusp, bordered by four somewhat similar cusped laterals.

→

FIGURES 81-86. Details of the radula of *Turcica panamensis*, new species. The radula of *Turcica*, like that of most other rhipidoglossate snails, is complex and best shown photographically divided into segments, as each individual element, especially amongst the marginals, appears differently according to the manner in which it is seen: 81, a freed section of the marginals composed of three or more complete sets, showing its bushy form.—82, 83, Magnified extremities of the outer set of filaments, above and below.—84, A single booklet from the inner side of the marginals, showing the broadly leaflike shape of the inner filaments and narrower shape of the outer ones.—85, Outer filaments of booklet illustrated in Figure 84, showing their narrow, bladelike shape and serrated edges.—86, A row of lateral and central teeth from the middle zone of ribbon.

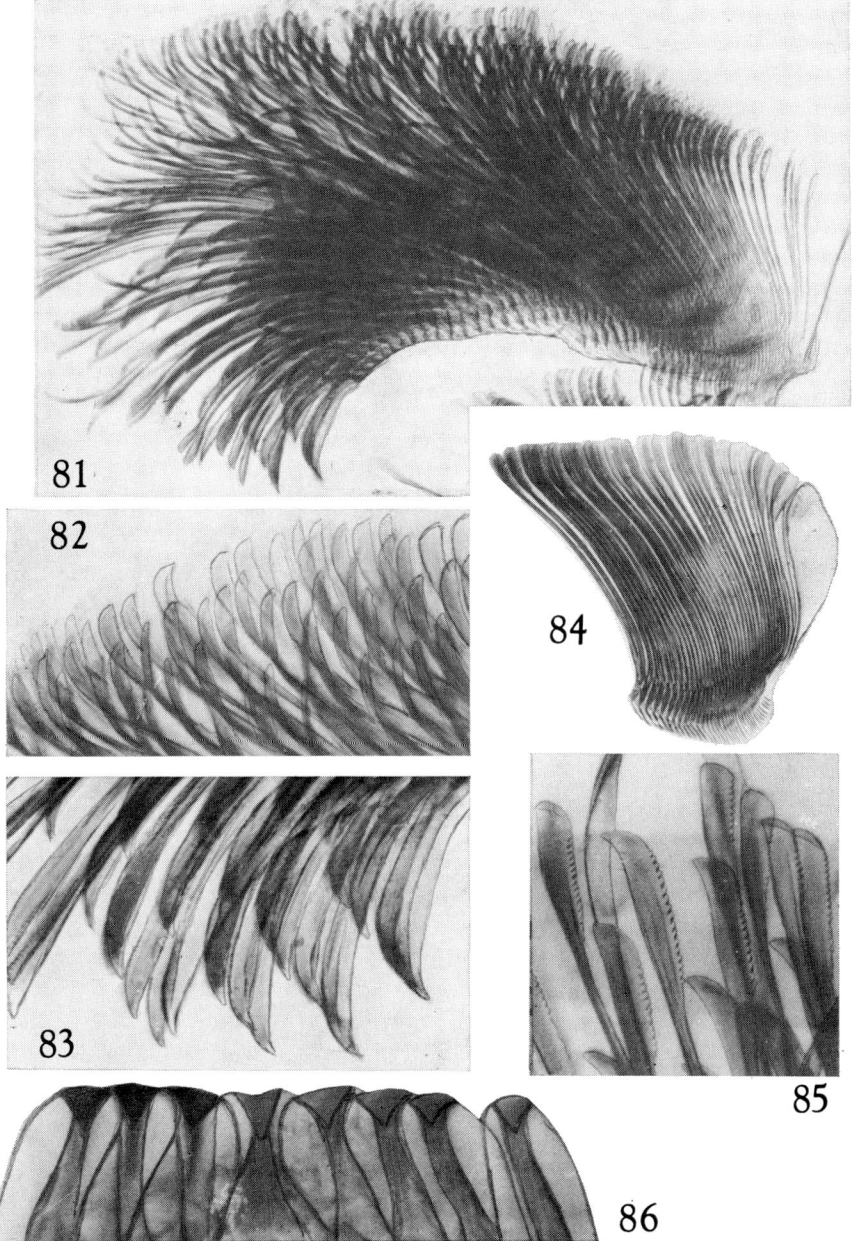

Turcica panamensis, new species
Figs. 75, 76, 81-86

Description.—The shell is of moderate size for the genus, thin, imperforate, and of a brownish olivaceous color; the interior is pearly. The spire is elevated, conic. The profile of each whorl is obliquely convex, with a prominent noded peripheral spiral cord which slightly overhangs the excavated sutural zone. Sculpture is produced by neatly beaded spiral cords, whose distribution is shown by the figure, with the largest cord being the peripheral one, the second largest an interval above. The spiral cords on the base are smaller and more finely beaded. The columellar area is excavated and bears a double tooth on the pillar.

Measurements.—Length 31.0 mm, diameter 29.0 mm, holotype; length 22.2 mm, diameter 19.0 mm, paratype.

Holotype.—USNM No. 701214, PILLSBURY Sta. P-529.

FIGURES 87-94. Turrid radulae: 87, *Drillia rosolina* Marrat, West Africa; the radular ribbon of a typical species of *Drillia*, showing the characteristic pattern of all genera of the Clavinae. Teeth are free and distinct and form a quinqueserial pattern. Also characteristic is the half-gear shape of the lateral tooth and the small central tooth. The formula is 1.1.1.1.1.—88, *Hormospira maculosa* (Sowerby), PILLSBURY Sta. P-502. In ribboned turrids, other than those of the Clavinae, the lateral and marginal teeth are combined so as to resemble and function as a single tooth, but they often become separated during extraction (as in Fig. 89). In *Hormospira*, the central tooth is well developed with a high, sharp medial cusp. Although the pattern seems to be triserial, it is actually quinqueserial, and its formula may be written as $1/1.1.1\backslash1$, the inclined line showing that the outer marginal element overlies the lateral.—89, *Knefastia? pilsbryi* (Lowe), PILLSBURY Sta. P-502; a ribboned turrid in which the central zone of the radula is open and empty, lacking the central row of teeth. The partial and full displacement of the marginal teeth overlying the lateral is shown. Formula may be written as $1/1.0.1\backslash1$.—90, *Steiraxis aulaca* Dall, PILLSBURY Sta. P-526; a ribboned turrid in which the central teeth are small. Marginal teeth are shown as partly overlapping the laterals; formula same as for Figure 88.—91, *Brachytoma stromboides* (Sowerby), PILLSBURY Sta. P-529; a toxoglossate radula. Line of bubbles within shows that the shaft of the tooth is hollow; length of tooth 1.9 mm.—92, *Carinodrillia dariena*, new species, PILLSBURY Sta. P-493; a partial toxoglossate radula in which the teeth are shown in their normal, biserial arrangement, attached by their rounded bases to two strings (as in *Conus*). Each tooth is harpoon-shaped, with two hooked barbs, one on each side behind the needle-sharp point. Each tooth measures approximately 1.9 mm.—93, *Cruziturricula panthea* (Dall), PILLSBURY Sta. P-553; toxoglossate. Note shape of tooth with shank, hooked barb near end, and hooked blade on opposite side; also shape of base. Individual tooth about 0.45 mm in length.—94, *Carinodrillia duplicata* (Sowerby), PILLSBURY Sta. P-543; toxoglossate. A small string of teeth; length of an individual tooth is 0.20 mm.

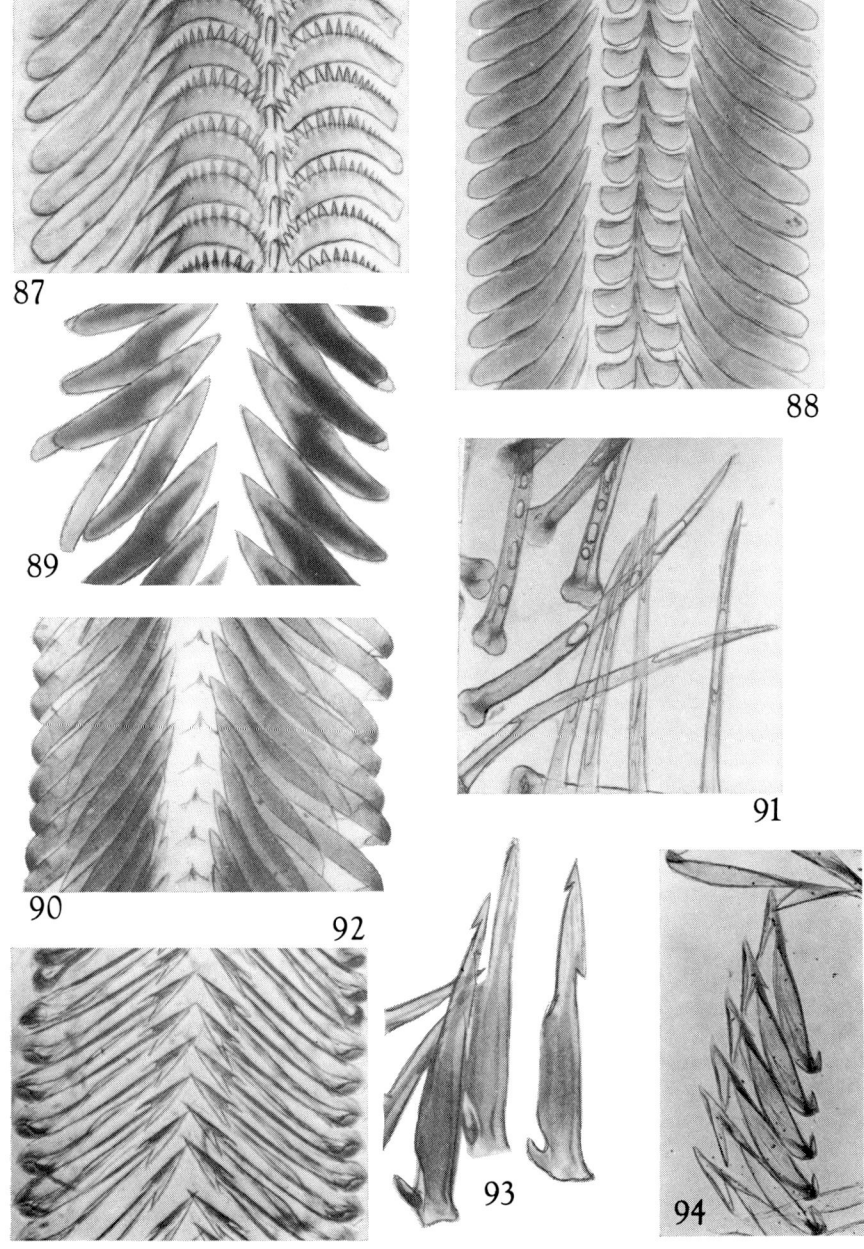

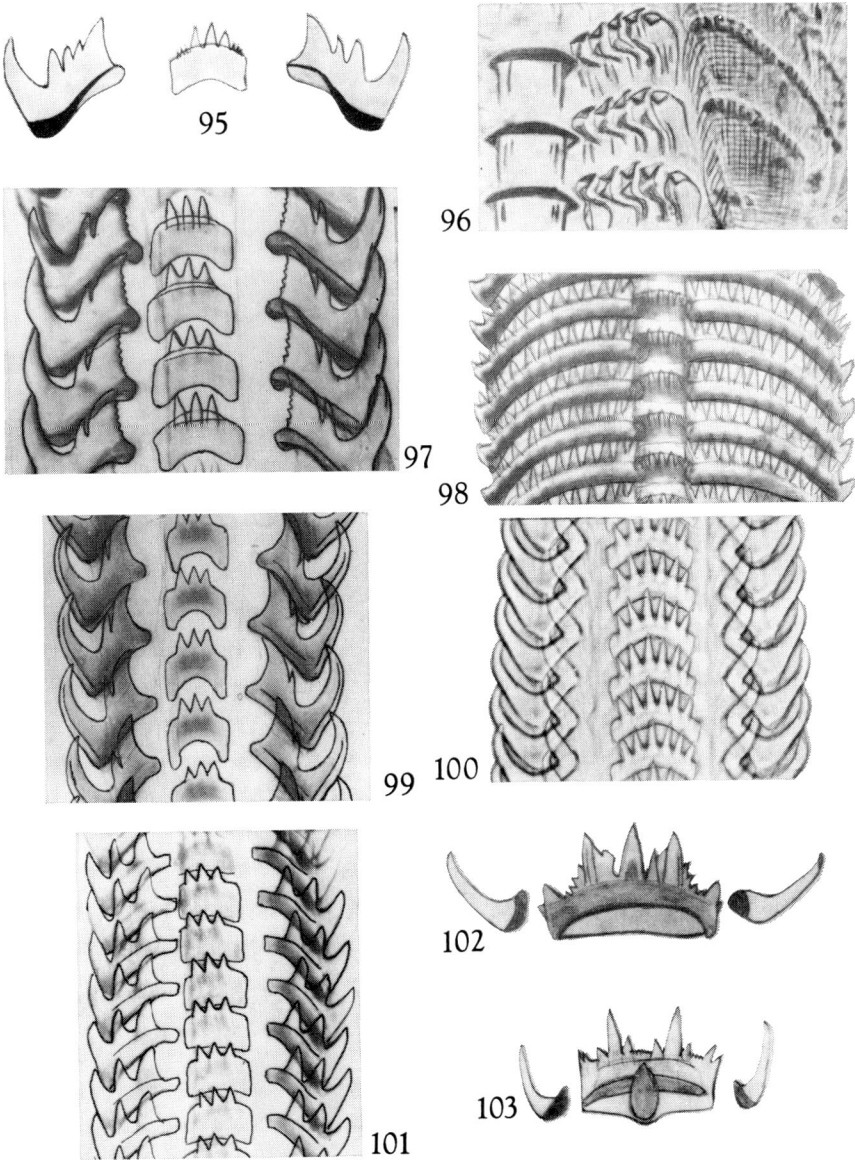

FIGURES 95-103. Gastropod radulae: 95, *Cantharus tranquebaricus* Gmelin; sketch of a single row of teeth. Specimen trawled in 3-15 fathoms, Porto Novo, S. Arcot Dist., Madras State, India; A. J. Kohn, 1969. Length of ribbon 13.3 mm, open width 0.46 mm; about 144 rows of teeth. USNM.—96, *Panocochlea*

GRAY, JOHN EDWARD
1824. Monograph of the Cypraeidae. Zool. Jour., *1:* 71-80; 137-152; 367-391.
HERTLEIN, LEO GEORGE AND A. M. STRONG
1940. Eastern Pacific Expeditions of the New York Zoological Society. XXII. Mollusks from the west coast of Mexico and Central America. Part 1. Zoologica, N.Y., *25*(4): 369-430, pls. 1-2.
1951. Eastern Pacific Expeditions of the New York Zoological Society. XLIII. Mollusks from the west coast of Mexico and Central America. Part X. Zoologica, N.Y., *36*(2): 67-120, pls. 1-11.
HIDALGO, JOAQUÍN GONZÁLEZ
1906. Obras malacológicas. Monografía de las especies vivientes del género *Cypraea.* Mems. R. Acad. Cienc. Madrid, *25:* xv + 588 pp.
HINDS, RICHARD BRINSLEY
1844. The zoology of the voyage of H.M.S. Sulphur, under the command of Captain Sir Edward Belcher . . . during . . . 1836-42. Edited and superintended by Richard Brinsley Hinds, Esq., Surgeon, R.N., attached to the expedition. Vol. II. Mollusca. Smith, Elder and Co., London, pp. 1-72 + i-[v], pls. 1-21.
LAMARCK, J. B. P. A.
1810. Suite de la détermination des espèces de mollusques testacés. Continuation du genre Porcellaine et des genres Ovula, Tarrière, Ancillaire, et Olive. Annls Mus. Hist. nat. Paris, *16:* 89-114; 300-328.
LOWE, H. N.
1935. New marine Mollusca from west Mexico, together with a list of shells collected at Punta Peñasco, Sonora, Mexico. Trans. S Diego Soc. nat. Hist., *8*(6): 15-32, pls. 1-4.
MCLEAN, JAMES H.
1970. Notes on the deep water Calliostomas of the Panamic Province with descriptions of six new species. Veliger, *12*(4): 421-426, pl. 62.
OLSSON, AXEL A.
1961. Mollusks of the tropical eastern Pacific, particularly from the southern half of the Panamic-Pacific faunal province (Panama to Peru). Panamic-Pacific Pelecypoda. Paleontological Research Institution, Ithaca, N.Y., 574 pp., 86 pls.
1964. Neogene mollusks from northwestern Ecuador. Paleontological Research Institution, Ithaca, N.Y., 256 pp., 38 pls.
PETIT DE LA SAUSSAYE, SAUVEUR
1852. Descriptions de coquilles nouvelles. J. Conch., Paris, *3:* 51-59, pl. 2.
PILSBRY, HENRY AUGUSTUS
1889. Trochidae, Stomatiidae, Pleurotomariidae, Haliotidae. Manual of Conchology, Vol. 11, 519 pp., 67 pls.
PILSBRY, HENRY AUGUSTUS AND H. N. LOWE
1932. West Mexican and Central American mollusks collected by H. N. Lowe. Proc. Acad. nat. Sci. Philad., *84:* 33-144, pls. 1-17.
PILSBRY, HENRY AUGUSTUS AND AXEL A. OLSSON
1935. New molluscs from the Panamic Province. Nautilus, *48*(4): 116-121, pl. 6.
REEVE, LOVELL AUGUSTUS
1843- Monograph of the genus *Pleurotoma.* Conchologia Iconica, *1:* 40
1846. pls., [80] pp.

SCHILDER, F. A.
- 1933. Beiträge zur Kenntnis der Cypraeacea. VI. Zool. Anz., *101:* 180-193.

SOWERBY, GEORGE BRETTINGHAM (FIRST OF THE NAME)
- 1832. *In:* Sowerby, James. The genera of Recent and fossil shells, for the use of students in conchology and geology. Commenced by J. Sowerby . . . and continued by G. B. Sowerby . . . with . . . plates by J. Sowerby and J. D. C. Sowerby. 2 vols., 267 pls. with descriptive letterpress. London, [1820-25]-34.
- 1833. Characters of new species of Mollusca and Conchifera, collected by Hugh Cuming. Proc. zool. Soc. Lond., for 1832: 194-202. (Published 1833.)

SOWERBY, GEORGE BRETTINGHAM, JR.
- 1832-1841. The conchological illustrations. London, iv + 116 pp., 200 pls.
- 1870. Thesaurus conchyliorum, or monographs of genera of shells. Vol. 4, parts 26-28, *Cypraea,* pp. 1-58, pls. 292-328.
- 1879. Thesaurus conchyliorum, or monographs of genera of shells. Vol. 4, parts 33-34, *Murex,* pp. 1-55, pls. 380-403.

THIELE, JOHANNES
- 1925. Gastropoda der deutschen Tiefsee-Expedition. II. Teil. Wiss. Ergebn. dt Tiefsee-Exped. 'Valdivia,' *17*(2): 35-382, pls. 13-46.
- 1929-1931. Handbuch der systematischen Weichtierkunde. Vol. 1. Jena, Verlag von Gustav Fischer, viii + 778 pp.

TROSCHEL, FRANZ HERMANN
- 1856-1893. Das Gebiss der Schnecken zur Begründung einer natürlichen Classification untersucht von . . . F. H. Troschel (fortgesetzt von . . . J. Thiele). 2 vols. Berlin.

WENZ, WILHELM
- 1938-1944. Gastropoda. Allgemeiner Teil und Prosobranchia. *In* O. H. Schindewolf (Ed.), Handbuch der Paläozoologie, 6: 1-1639.

WOODRING, WENDELL PHILLIPS
- 1928. Contributions to the geology and paleontology of the West Indies. Miocene Mollusks from Bowden, Jamaica. Part 2. Publs Carnegie Instn, No. 385, vii + 564 pp., 40 pls.

Biological Results of the University of Miami Deep-Sea Expeditions. 78.

THE CONIDAE OF THE PILLSBURY EXPEDITION TO THE GULF OF PANAMA[1]

JAMES NYBAKKEN

Moss Landing Marine Laboratories, Box 223, Moss Landing, California
and
Department of Zoology, University of Washington, Seattle, Washington

ABSTRACT

Four hundred and forty-nine specimens of 11 species of the molluscan genus *Conus* were obtained by R/V JOHN ELLIOTT PILLSBURY expedition to the Gulf of Panama in 1967. Of these, 446 specimens and 10 species were obtained by dredge and trawl in depths between 15 and 210 meters. The literature on subtidal cones is reviewed, the ten species are discussed and in some cases the descriptions amplified, and *Conus poormani* Berry is figured for the first time. All ten species are illustrated, and the faunal relationships of *Conus* on the two sides of the Isthmus of Panama are discussed.

INTRODUCTION

The collection of *Conus* on which the present paper is based was obtained by the R/V JOHN ELLIOTT PILLSBURY during a biological survey of the Gulf of Panama, April 29–May 11, 1967. The expedition was under the direction of Dr. Gilbert Voss.

The cruise track of the ship covered the major portion of the gulf (see Fig. 1). A total of 197 stations was occupied, of which 37 produced one or more specimens of *Conus* (Fig. 1).

With the exception of one specimen taken with a box dredge and two taken at two shore stations near Balboa, all specimens of *Conus* were taken in a 10-foot otter trawl at depths of 15 to 210 meters. A total of 449 individuals of 11 species of *Conus* was obtained. This paper reports only on these subtidal species of *Conus*, which explains the absence here of many of the more common shallow-water and intertidal species of the West American Conidae. I have also omitted the two specimens from Balboa, as they were worn beach specimens; this removes one species from consideration, leaving ten discussed here.

The itinerary of the expedition and a full description of the stations occupied and fauna collected has been reported by Voss (1967).

I am indebted to Dr. Gilbert Voss for making this collection available to me for study. I wish to acknowledge also the valuable assistance and advice of Dr. G. Dallas Hanna, Allyn Smith, and L. G. Hertlein of the California Academy of Sciences, who made available to me the Academy's

---

[1] This paper is one of a series resulting from the National Geographic Society–University of Miami Deep-Sea Biology Program. The work was supported in part by National Science Foundation grant GB-5776 and the National Geographic Society–University of Miami Deep-Sea Biology Program.

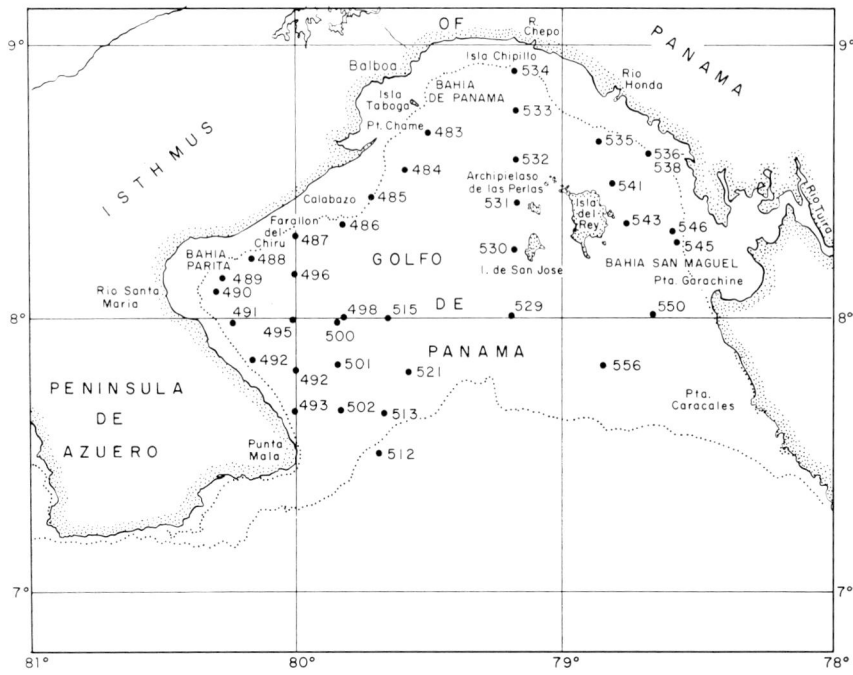

FIGURE 1. The Gulf of Panama. The numbers indicate station numbers at which *Conus* was taken.

collections for verification of identifications. I am also grateful to Dr. James McLean of the Los Angeles County Museum for his loan of specimens of *Conus poormani* and for photos of the type of *C. poormani*. I am especially grateful to Dr. Alan Kohn for his advice, suggestions, and help in working over this collection, as well as for making available to me his collection of type photographs of species of *Conus* and his extensive collection of literature on *Conus*. Finally, I would like to acknowledge the support received from NSF grant GB-5942X during the preparation of this manuscript.

LITERATURE REVIEW

Since the West American species of the family Conidae were first summarized by Dall (1910), they have been the subject of an extensive recent review by Hanna (1963), which brought up to date the earlier review by Hanna & Strong (1949). Emerson & Old (1962) reported on the Conidae of the PURITAN Expedition to the Gulf of California and west Mexico. The reader is referred to these reviews for complete synonymies of the species considered here. One more recently published synonym is cited here.

The early literature concerning the marine molluscan fauna of the Gulf of Panama area has been reviewed by Hertlein & Strong (1955) and Strong & Hertlein (1939). The latter paper gives a particularly fine overview. Keen's (1958) book covers most species of shelled mollusks of the extensive Panamic province and briefly reviews the history of malacology in the area.

Despite the recent reviews and the relatively great amount of knowledge available on the West American Conidae, a considerable amount of uncertainty still exists concerning the exact number of species present. This is a general situation for *Conus*, as Von Mol, Tursch & Kempf (1967) have pointed out, due in great part to a lack of knowledge of the intraspecific variability. Since all species of *Conus* have been established only on the bases of shell morphology and color pattern, individuals or groups with deviant color patterns or slight changes in form have often been considered as new species. The animal, its ecology, or internal anatomy have rarely been the subject of study. Kohn (1959, 1968) has recently done much to rectify this lack of knowledge of the animals, and I hope that a study of mine (Nybakken, 1970b) on the radular structure of the West American species will further help to clarify some of the systematic problems. All identifications in this paper have been checked against the radular structure, where the radula has been shown to be useful.

Although some malacologists have attempted to assign West American species of the genus *Conus* to several subgenera based on shell characters (Robertson, 1954; Keen, 1958), I have refrained from this practice and have followed Hanna (1963) in retaining all species in the single genus *Conus*. Emerson & Old (1962) have given an exhaustive account of the supraspecific taxa which have been proposed and used in *Conus*.

All photographs here are of the shells of animals collected alive, which still retain the periostracum. This has obscured the color pattern in a few cases, and in these cases I have included photographs of specimens both with and without periostracum.

Conus archon Broderip
Fig. 2

Conus archon Broderip, 1833, Proc. Zool. Soc. Lond., May 24: 54.

Collecting Stations.—P-493, 7°39.5'N, 80°0.7'W, 33-37 meters, bottom of shell rubble, 1 dead and 2 live specimens.

Range.—Gulf of California to Panama (Hanna); "Mazatlan, Acapulco, west coast of Central America" (Dall).

Remarks.—This does not appear to be a common species, even in dredged material, and apparently does not occur intertidally or in shallow water (Keen, 1958). Only three specimens are represented in the present collection, of which two were taken alive. Emerson & Old (1962) did not obtain this species in their collecting in the Gulf of California, nor did I, during two and one-half months of similar collecting in the same area in 1967.

This is a readily recognizable species. Figure 2 shows the larger of the two living specimens obtained.

Conus arcuatus Broderip & Sowerby
Fig. 3

Conus arcuatus Broderip & Sowerby, 1829, Zool. Jour., *4*(15): 379, Jan. 1829.

Collecting Stations.—P-483, 8°40.5'N, 79°30.7'W, 22 meters, 12 live and 5 dead specimens.—P-484, 8°33.2'N, 79°35.4'W, 31 meters, 1 live and 2 dead specimens.—P-485, 9°26.2'N, 79°43.2'W, 15 meters, 3 live and 4 dead specimens.—P-486, 8°20'N, 79°49.7'W, 20 meters, 5 dead shells.—P-487, 8°18.1'N, 80°0.5'W, 18 meters, 1 live and 8 dead shells.—P-488, 8°13.1'N, 80°9.6'W, 16 meters, 6 dead specimens.—P-490, 8°6.3'N, 80°18.2'W, 22 meters, 3 dead shells.—P-491, 7°59.1'N, 80°14.2'W, 20 meters, 6 dead shells.—P-492, 7°50.7'N, 80°9.8'W, 18 meters, 9 dead specimens.—P-493, 7°39.5'N, 80°0.7'W, 36 meters, 16 live and 20 dead specimens.—P-494, 7°49.3'N, 80°00'W, 45 meters, 8 live and 3 dead specimens.—P-495, 7°59.2'N, 80°0.2'W, 40 meters, 4 live and 9 dead specimens.—P-496, 8°9.7'N, 80°0.9'W, 33 meters, 4 live and 11 dead specimens.—P-498, 8°00'N, 79°49.8'W, 57 meters, 1 dead shell.—P-500, 7°59.7'N, 79°49.7'W, 53 meters, 5 dead shells.—P-502, 7°40'N, 79°50.5'W, 79 meters, 1 dead shell.—P-515, 8°0.4'N, 79°40.8'W, 78 meters, 1 dead shell.—P-529, 8°0.7'N, 79°11.8'W, 84 meters, 1 dead shell.—P-530, 8°15.3'N, 79°10.6'W, 60 meters, 1 live and 4 dead specimens.—P-531, 8°25.5'N, 79°10.7'W, 57 meters, 7 live and 4 dead specimens.—P-532, 8°35.4'N, 79°10.5'W, 69 meters, 13 live and 27 dead specimens.—P-533, 8°45.2'N, 79°10.3'W, 36 meters, 2 live and 6 dead specimens.—P-534, 8°54.5'N, 79°10.2'W, 16 meters, 1 dead shell.—P-535, 8°38.6'N, 78°51.9'W, 31 meters, 1 live and 10 dead specimens.—P-536, 8°35.6'N, 78°40.7'W, 18 meters, 1 live and 1 dead specimen.—P-541, 8°30'N, 78°49.2'W, 29 meters, 2 dead shells.—P-543, 8°21.4'N, 78°45.5'W, 62 meters, 1 dead shell.—P-545, 8°16.7'N, 78°34.3'W, 24 meters, 1 dead shell.—P-546, 8°19.2'N, 78°35.8'W, 28 meters, 3 live and 5 dead specimens.

Range.—Cedros Island, Baja California (Dall), Gulf of California south to Panama (Hanna).

Remarks.—This species was by far the most common, both in terms of individuals and number of stations at which it was found. Of the 449 specimens collected, 238 were *C. arcuatus*. This species is dominant enough to have been considered by Parker (1964) as an indicator species of certain medium-depth shelf areas in the Gulf of California. Certainly the results of this expedition indicate that it is also the dominant species of *Conus* in depths of 15-75 meters in the Gulf of Panama. This suggests that it is probably common at these depths, wherever substrates are suitable, at all intermediate localities along the rest of Central America.

In light of the above, it is strange that Emerson & Old (1962) did not obtain the species in their extensive dredging in the Gulf of California, and that Hertlein & Strong (1955) report only three specimens from two stations in the Gulf of Panama.

The species is readily recognizable and not easily confused with any other West American species of *Conus*.

Despite the abundance of this species in dredging, it seems never to be taken in depths above 10 meters or in the intertidal zone.

Conus fergusoni Sowerby
Figs. 4, 5

Conus fergusoni G. B. Sowerby III, 1873, Proc. Zool. Soc. Lond.: 145, pl. 15, fig. 1.
Conus chrysocestus Berry, 1968, Leaflets in Malacology, *1*(25): 157.

Collecting Stations.—P-546, 8°19.2'N, 78°35.8'W, 28 meters, mud and shell bottom, 1 dead specimen.—P-550, 8°0.7'N, 78°40.3'W, 64 meters, pebble bottom, 1 live specimen.

Range.—West coast of Baja California (Hanna), Gulf of California as far north as San Pedro Bay, Sonora (Emerson & Puffer), and south to Mancora, Peru and the Galapagos Islands (Hanna).

Remarks.—As Hanna (1963) has noted, the young of this species differs from the adult, both in shape and in coloration. The large adults are usually white, whereas the young are often a bright yellow with one or two white bands on the body whorl. The young were originally described by Dall as *Conus xanthicus*. Emerson & Old (1962) illustrated some of these juvenile shells (Figs. 13, 14, 15), as did Hanna (1963), who illustrated them in color. The single live specimen in this collection closely resembles Emerson and Old's Figure 14.

It is possible that Dall's *C. xanthicus* is, in fact, a valid species. In the course of a study of the radular structure of the West American *Conus* (Nybakken, 1970b), I found that whereas certain of the small juvenile forms had radular teeth typical of *C. fergusoni*, others did not. In fact, the teeth of certain individuals were quite different and approached the teeth of *Conus regularis* in morphology. This suggests that either there are two species involved, one of which is the juvenile *C. fergusoni* and the

FIGURES 2-13. Conidae from the Gulf of Panama: 2, *Conus archon* Broderip; shell length 41.3 mm, Sta. P-493.—3, *Conus arcuatus* Broderip & Sowerby; shell lengths from left to right are 32.5 mm, 36.7 mm, 38.3 mm, and 39.6 mm; Sta. P-432.—4, *Conus fergusoni* Sowerby; shell length 51.8 mm, Sta. P-550.— 5, *Conus fergusoni* (?); shell length 40.5 mm, Sta. P-550.—6, *Conus gradatus* Mawe; shell length 61.1 mm, Sta. P-546.—7, *Conus patricius* Hinds; shell length 80.0 mm, Sta. P-496.—8-12, *Conus poormani* Berry: 8, apertural view of shell

other presumably *C. xanthicus* or an undescribed form, or that there is a change in tooth morphology with age in *C. fergusoni*. If the latter case proves to be true, it will be the first time that this has been recorded for *Conus*. At present, insufficient data are available to make any decision. The present live specimen has typical *C. fergusoni* teeth of the adult type.

The present collection also contained another specimen collected alive at station P-550, which originally was identified as *C. fergusoni* (Fig. 5). Subsequent analysis of its radular teeth showed that their morphology was quite different from that of the teeth of typical *C. fergusoni* and similar to that of the teeth of *C. regularis*. In light of the problem with the radula, as outlined above, I am retaining it here as *C. fergusoni*, though it may well prove not to be this species. The shell resembles the specimen identified as *C. fergusoni* and illustrated in figure 15 of Emerson & Old (1962), as well as their description of it.

I have also seen specimens of *Conus chrysocestus* Berry, 1968, which, unfortunately, lacked the soft parts. They are similar to young shells of *C. fergusoni* and seem best considered as such, pending further study. They were not taken on this expedition.

Conus gradatus Mawe
Fig. 6

Conus gradatus Mawe, 1823, Linn. Syst. Conch.: 90.

Collecting Stations.—P-521, 7°48.3′N, 79°35.1′W, 119 meters, 1 dead specimen.—P-546, 8°19.2′N, 78°35.8′W, 28 meters, 1 live specimen.

Range.—Cedros Island, west coast of Baja California (Keen), Gulf of California as far north as Mulege (Hanna), south to Panama (PILLSBURY Expedition) and Clipperton Island (Hanna).

Remarks.—I have assigned only two PILLSBURY specimens to this species, referring all other questionable individuals to *C. regularis*. The two specimens here designated as *C. gradatus* are quite different in shell morphology from the typical *C. regularis* (compare Figs. 6 and 18), and hence their separation. Both individuals are very large (61.1 and 61.3 mm) and have rounded shoulders with high spires. The body whorl is tapered and varies from slightly concave in outline near the anterior end to slightly convex

---

←

with periostracum; shell length 44.8 mm, Sta. P-531; 9, dorsal view of same shell as in Figure 8; 10, apertural view of shell without periostracum; shell length 46.5 mm, Sta. P-531; 11, dorsal view of same shell as in Figure 10, 12, series of juveniles showing variations in color pattern; shell lengths from left to right are 32.6 mm, 32.1 mm, 28.3 mm, 26.7 mm, and 24.2 mm; Sta. P-529.—13, *Conus recurvus* Broderip; shell length 50.2 mm, Sta. P-529.

near the shoulder. The outer lip is incurved and approaches that of *C. recurvus*. The color pattern is similar to that seen in specimens of either *C. recurvus* or *C. regularis*. The periostracum of the live specimen is brown and much heavier than the periostracum of any of the specimens of *C. regularis* or *C. recurvus* in this collection. The radular teeth are typical of the *C. regularis* species group. Figure 6 illustrates the single live specimen.

Conus patricius Hinds
Fig. 7

Conus patricius Hinds, 1843, Ann. Mag. Nat. Hist., N. S., *11*(70): 256.

Collecting Stations.—P-496, 8°9.7′N, 80°0.9′W, 33 meters, 2 live specimens.—P-534, 8°54.5′N, 79°10.2′W, 16 meters, mud bottom, 1 juvenile.

Range.—Acapulco, Mexico (Emerson & Hertlein) south to Ecuador (Keen) and the Galapagos Islands (Dall).

Remarks.—The two large specimens showed the typical deep brown periostracum covering an otherwise white and very heavy shell. The species does not range into the Gulf of California and was not obtained by Emerson & Old (1962). Figure 7 illustrates one of the two mature individuals.

Conus poormani Berry
Figs. 8, 9, 10, 11, 12

Conus poormani Berry, 1968, Leaflets in Malacology: *1*(25): 156.

Collecting Stations.—P-529, 8°0.7′N, 79°11.8′W, 84 meters, 1 dead and 18 live specimens.—P-530, 8°15.3′N, 79°10.6′W, 60 meters, hard bottom, 1 live specimen.—P-531, 8°25.5′N, 79°10.7′W, 57 meters, hard bottom, 5 live specimens.

Range.—Gulf of California, 24-26 fathoms, off Morro Colorado, Sonora, Mexico (type-locality), to the Gulf of Panama (PILLSBURY Expedition).

Remarks.—This species was originally described without a figure, but, thanks to Dr. James McLean of the Los Angeles County Museum, I have been able to examine photographs of the type-specimen as well as other specimens from their collection. The present material matches the Los Angeles County Museum specimens and the original description in all respects.

I have chosen to figure several specimens here, to compensate for the lack of an illustration with the original description, to facilitate identification by other malacologists, and to show the range of variation as far as now known.

I have not searched the 3000 or so names in the systematic literature of *Conus* to see if Berry's name should be replaced by an earlier available name, and until that is done it seems prudent to let his name stand. It does seem strange, however, that so distinctive and relatively large a species as this one should have remained undescribed until 1968.

Berry (1968) stated that the most distinctive feature of *C. poormani* is the curious "fringe" exhibited by the periostracum on the shoulder (Figs. 8, 9, 12). He considered this feature unique among the West American species of *Conus*. However, the present collection has specimens of *C. virgatus* which show the same fringe on the shoulder (Fig. 23). *Conus poormani* can be readily separated from *C. virgatus* by the angulated shoulders, as opposed to the rounded shoulders of *C. virgatus*, and by the pattern of brown markings on the shell. These occur in irregular broken flammules or as blotches in *C. poormani*, but do not form long axial stripes as in *C. virgatus*. Some specimens of *C. poormani* have a median white stripe separating the brown markings (Fig. 10), which is not found in *C. virgatus*. The color pattern in some specimens of *C. poormani* is very similar to that of typical *C. recurvus* (see Fig. 10), which it also resembles in general shell morphology (Fig. 10). It may be distinguished from *C. recurvus* by the aforementioned peculiar periostracal fringe and by the difference in radular teeth.

The radular teeth of *C. poormani* are similar to the teeth of *C. virgatus* and the species of the *C. regularis* complex. I have described and illustrated them elsewhere (Nybakken, 1970b). They are thus very different from the teeth of *C. recurvus*, and this difference indicates that *C. poormani* is not a color variant of *C. recurvus*.

The small individuals of *C. poormani* in the present collection differ somewhat in color pattern from the larger individuals, in that the brown is distributed in larger patches which then accentuate the median white stripe (Fig. 12). In this respect, the small ones begin to approach the young of *C. fergusoni* and could be confused with *C. fergusoni*; however, the periostracal fringe should readily separate the two.

From the station data it would appear that this species is restricted to hard bottoms, which may account for its seemingly patchy distribution. It is a species which appears to be restricted to subtidal areas.

Conus recurvus Broderip
Figs. 13, 14, 15, 16, 17

Conus recurvus Broderip, 1833, Proc. Zool. Soc. Lond.: *1*(4): 54.

Collecting Stations.—P-487, 8°18.1′N, 80°0.5′W, 18 meters, 1 live specimen.—P-491, 7°59.1′N, 80°14.2′W, 20 meters, 1 dead shell.—P-493, 7°39.5′N, 80°0.7′W, 36 meters, 2 live and 2 dead specimens.—P-494, 7°49.3′N, 80°00′W, 45 meters, 1 dead and 3 live specimens.—P-495,

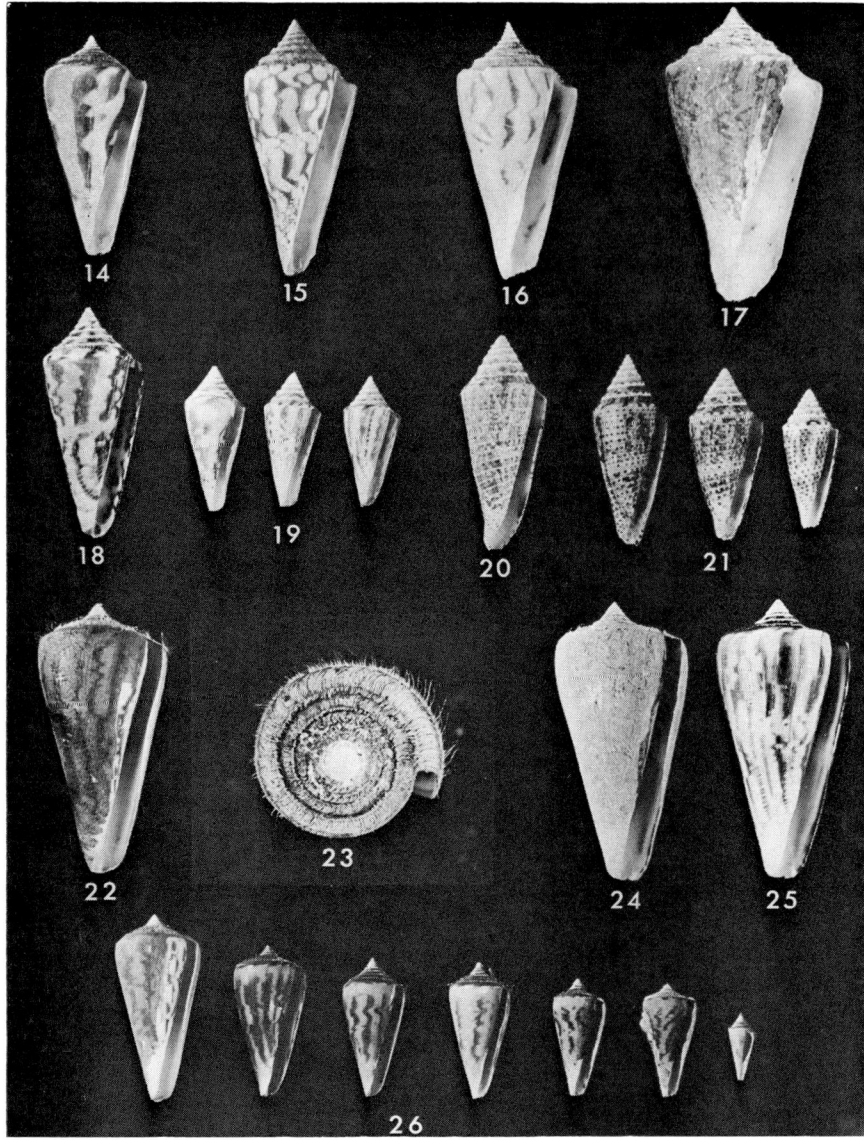

FIGURES 14-26. Conidae from the Gulf of Panama: 14-17, *Conus recurvus* Broderip: 14, shell length 47.5 mm, Sta. P-531; 15, shell length 50.0 mm, Sta. P-531; 16, specimen from 117 meters, shell length 54.2 mm, Sta. P-513; 17, specimen from 210 meters, shell length 68.6 mm, Sta. P-512.—18-19, *Conus regularis* Sowerby: 18, shell length 42.5 mm, Sta. P-492; 19, shell lengths from

7°59.2′N, 80°0.2′W, 40 meters, 2 dead shells.—P-500, 7°59.7′N, 79°49.7′W, 53 meters, 1 live specimen and 2 dead shells.—P-501, 7°50.2′N, 79°50.5′W, 68 meters, 2 dead shells.—P-512, 7°30.5′N, 79°41.5′W, 210 meters, 2 live specimens.—P-513, 7°39.5′N, 79°40.7′W, 117 meters, 1 live specimen.—P-529, 8°0.7′N, 79°11.8′W, 84 meters, 1 live and 2 dead specimens.—P-531, 8°25.5′N, 79°10.7′W, 57 meters, 1 dead and 5 live specimens.

Range.—Outer coast of Baja California as far north as Magdalena Bay (Hanna), the Gulf of California as far north as Los Angeles Bay (McLean); southward to Ecuador (Emerson & Old).

Remarks.—This species has often been included with the *Conus regularis-scalaris-gradatus* complex due to the difficulty of separating certain shells from that complex (see Hanna [1963] for a discussion). Hanna feels that this species includes such a wide range of variations in color pattern, and even shell shapes, that it is difficult to assign individuals to the species. That a variation in color pattern in particular exists in this species cannot be denied, and I have illustrated the patterns of the specimens from this expedition in Figures 13 to 17. However, I have demonstrated in recent work (Nybakken, 1970b) that the radular teeth of *C. recurvus* are quite different from those of all species of the *C. regularis* complex; if live collected material is available it is easy to assign individuals to *C. recurvus* or to the *C. regularis* complex of species. The difference in structure of the teeth also suggests that, despite the sometimes similar shell coloration and shape, *C. recurvus* is, in fact, not close to *C. regularis* and certainly exploits a different food source. Hence, I cannot agree with Hanna that this species should be united with the *C. regularis* complex.

The radular teeth of all specimens illustrated in Figures 13 to 17 are similar and typical of *C. recurvus.* The range of variation in color pattern is apparent here. It is of some interest to note that the specimens with the least amount of brown marking come from the deepest water. The

---

←

left to right: 28.6 mm, 26.5 mm, and 26.0 mm; Sta. P-492.—20-21, *Conus tornatus* Sowerby: 20, shell length 44.8 mm, Sta. P-487; 21, shell lengths from right to left are 27.0 mm, 33.3 mm, and 37.2 mm; Sta. P-489.—22-26, *Conus virgatus* Reeve: 22, specimen illustrating how the color pattern shows through a wet, intact periostracum; shell length 51.8 mm, Sta. P-535; 23, view of the spire of same specimen as in Figure 22, showing the "fringe" of periostracal hairs; 24, specimen with an intact, dry periostracum; shell length 48.9 mm, Sta. P-490; 25, same specimen as in Figure 24, but with the periostracum removed to show the color pattern; 26, growth series showing also some of the color patterns found in the species; shell lengths from right to left are 15.0 mm, 24.7 mm, 25.3 mm, 29.4 mm, 30.0 mm, 31.8 mm, and 51.4 mm; Sta. P-492.

specimen in Figure 17 was taken in 210 meters, a depth which exceeds the depth range of 154 meters given by Keen (1958) for the species.

Thus far, all individuals which have had teeth typical of *C. recurvus* have also shown a similar set of shell characteristics. These are a wide aperture, a concave outline of the spire, a markedly curved outer lip, and a pronounced thinness of the shell such that it is easily broken. The latter feature has been noted by Keen (1958) and is one of the most characteristic features of the shell of the species.

The constancy of structure of the radular teeth in specimens of varying color pattern suggests that this is a valid species, but one with considerable plasticity in color pattern. It now seems possible to correlate color and structure of the shell with structure of the radular teeth so that a better appreciation of the variation of shell color and structure within *C. recurvus* can be gained, and from this appreciation to establish more accurately the range of variation in the morphology of the shell, so that the shells alone can be distinguished from those of the *C. regularis* complex.

I agree with Hanna (1963) that, based on the illustrations, *C. scariphus* Dall, 1910, should be included in *C. recurvus*. However, a study of the illustration of the type of *C. magdalenensis* Bartsch & Rehder, 1939, leads me to agree with Emerson & Old (1962) that it is referable to *C. regularis*.

I have examined a color photograph of the type of *C. recurvus* Broderip, and it appears similar in all features to shells which are referable to the *C. regularis-gradatus* complex. It does not appear similar to typical *C. recurvus* from this expedition or to those figured by Hanna (1963), Emerson & Old (1962), and Keen (1958). However, I have not examined the types, or color photographs of the types, of *C. emarginatus* Reeve, 1844, *C. incurvus* Broderip, 1833, or *C. zebra* Lamarck, 1810, the next earliest available names, so I cannot say if one of these names should be applied to what we are now calling *C. recurvus*. Certainly Reeve's figure of *C. emarginatus* is more similar to what we are presently calling *C. recurvus* than is Broderip's type.

Conus regularis Sowerby
Figs. 18, 19

Conus regularis G. B. Sowerby I, 1833, *in* G. B. Sowerby II, The Conchological Illustrations: 2, pl. 29, fig. 29.

Collecting Stations.—P-490, 8°6.3'N, 80°18.2'W, 22 meters, 4 live and 4 dead specimens.—P-491, 7°59.1'N, 80°14.2'W, 20 meters, 1 live and 5 dead specimens.—P-492, 7°50.7'N, 80°9.8'W, 18 meters, 7 live and 2 dead specimens.—P-493, 7°39.5'N, 80°0.7'W, 36 meters, 1 dead shell.—P-531, 8°25.5'N, 79°10.7'W, 57 meters, 1 dead shell.

Range.—West coast of Baja California as far north as Magdalena Bay (Keen) throughout the Gulf of California (Hanna) and south to Peru (Dall).

Remarks.—As pointed out by many investigators (Keen, 1958; Emerson & Old, 1962; Hanna, 1963), this is a very variable species, which grades into types described as *C. gradatus* at one end of the series and *C. scalaris* at the other. It also has a considerable benthonic range from the intertidal to 100 meters (Keen, 1958).

Because of the variability and the lack of any clear-cut distinction on an individual basis between most specimens of *C. regularis* and *C. gradatus*, I have referred all specimens from this expedition, save two, to the former. The two were considerably different in shell morphology, especially in size, and seemed more typical of shells which have been referred to *C. gradatus*. I have discussed them under that species.

The radulas of *C. regularis*, *C. scalaris*, and *C. gradatus* are indistinguishable in general morphology and therefore there may be but a single species involved, whose variations in shell morphology have been designated as separate species. If this does prove to be the case, then *C. gradatus* Mawe, 1823, is the oldest available name, as Keen (1958) and Emerson & Old (1962) have noted.

Some of the specimens collected alive on this expedition approach the form called *C. monilifer* Sowerby, 1833 (*C. regularis*) in having an elevated spire (Fig. 19). Others are more typical of the described species (Fig. 18). None has a color pattern resembling the specimens of *C. recurvus* taken on this expedition.

Conus tornatus Sowerby
Figs. 20, 21

Conus tornatus G. B. Sowerby I, 1833, *in* G. B. Sowerby II, The Conchological Illustrations: 2, pl. 29, fig. 25.

Collecting Stations.—P-487, 8°18.1'N, 80°0.5'W, 18 meters, 1 live specimen.—P-488, 8°13.1'N, 80°9.6'W, 17 meters, 1 dead and 10 live specimens.—P-490, 8°6.3'N, 80°18.2'W, 18-22 meters, 2 live and 8 dead specimens.—P-491, 7°59.1'N, 80°14.2'W, 20 meters, 1 live specimen.—P-493, 7°39.5'N, 80°0.7'W, 36 meters, 9 dead shells.

Range.—Cedros Island on the west coast of Baja California (Dall), the Gulf of California to San Luis Island (Emerson & Old), and south to Ecuador (Hanna).

Remarks.—All specimens were taken on the west side of the gulf at stations bordering the Azuero Peninsula, in relatively shallow water, and were very large for the species. Keen (1958) gave the height of the species as 23 mm and Wagner & Abbott (1964) as 25 mm, but the one specimen from station P-487 was 44.8 mm in length. This is larger than the 36.4-mm specimen reported by Hertlein & Strong (1955) from Panama. Nine of the specimens from station P-488 exceeded 33 mm in height and five from station P-490 exceeded 30 mm, as did the one from station P-491 and eight of the nine from station P-493. The average height of all specimens taken was 31.1 mm, considerably above the size given by Keen (1958) and Wagner & Abbott (1964). Figure 20 illustrates the 44.8-mm specimen.

Conus tornatus appears to be a common subtidal species, as several authors have noted previously (Keen, 1958; Hanna, 1963; Emerson & Old, 1962).

Conus virgatus Reeve
Figs. 22, 23, 24, 25, 26

Conus virgatus Reeve, 1849, Conchologia Iconica, Suppl.: 1-2. (Name proposed for pl. 16, fig. 87, Conchologia Iconica, 1843.)

Collecting Stations.—P-487, 8°18.1′N, 80°0.5′W, 18 meters, 3 live specimens.—P-488, 8°13.1′N, 80°9.6′W, 16 meters, 3 live specimens.—P-490, 8°6.3′N, 80°18.2′W, 22 meters, 12 live and 4 dead specimens.—P-491, 7°59.1′N, 80°14.2′W, 20 meters, 2 live specimens.—P-492, 7°50.7′N, 80°9.8′W, 18 meters, 10 live and 2 dead specimens.—P-493, 7°39.5′N, 80°0.7′W, 36 meters, 12 live and 12 dead specimens.—P-494, 7°49.3′N, 80°00′W, 45 meters, 3 live specimens.—P-496, 8°9.7′N, 80°0.9′W, 33 meters, 1 live specimen.—P-533, 8°45.2′N, 79°10.3′W, 36 meters, 1 live specimen.—P-535, 8°38.6′N, 78°51.9′W, 31 meters, 3 live specimens.

Range.—West coast of Baja California as far north as Cedros Island, the Gulf of California as far north as Guaymas (Emerson & Old) and south to Ecuador (Keen).

Remarks.—After *C. arcuatus*, this species was the most abundant in the collection (70 specimens); most were taken at a few stations on the west side of the gulf bordering the Azuero Peninsula (Fig. 1).

Despite the variability in color pattern which can occur among individuals (Hanna, 1963; Hertlein & Strong, 1955), this is a distinctive species not readily confused with any of its West American congeners. The long tapered shell with the conspicuously rounded shoulder and low spire usually serve to identify the species even in worn beach shells. The live animal has a heavy brown periostracum through which the axial brown markings of the shell are visible only when the shell is wet (Fig. 22). The periostracum of some individuals projects as a fringe of "hairs" on the sutural lines of the spire (Fig. 23). The periostracum is thicker on older shells.

Hanna (1963) has indicated that there may be difficulty in distinguishing between the young of *C. fergusoni* and *C. virgatus*, but I do not consider this likely, especially if live-collected material is available. All sizes of *C. virgatus* studied here were readily distinguishable, and examination of both species in other collections has confirmed this. Figure 26 illustrates a series of growth stages from this collection and depicts both the variability in color pattern and the consistency of shell morphology from young to adult stages.

This species has been considered uncommon by Keen (1958), and Emerson & Old (1962) collected but four dead specimens in the Gulf of California. Hanna (1963) has further noted the absence of large numbers of specimens from recent expeditions. Hence this collection is of value for the large series of specimens of a particularly wide size-range.

The species is another which seems to be restricted to subtidal areas and is rarely encountered intertidally. It was most abundant in this expedition in depths of 16-40 meters.

Relationships of Atlantic and Pacific Species of *Conus*

Although the *Conus* faunas of the western Atlantic and the East Pacific are small in terms of number of species and relatively well known systematically, very little has been said concerning their possible relationships. Considering the closeness of the two faunas in a geographical sense, and the possibility that the two faunas may become confluent if a sea-level canal is constructed, it seems appropriate to mention here something about this relationship.

Ekman (1953) has pointed out that several invertebrate taxa and the shallow-water fish fauna of the East Pacific show much closer resemblance to the same groups in the Caribbean than they do to similar groups in the Indo-Pacific. Indeed, he points out that the East Pacific barrier is the most pronounced zoogeographical barrier for the circumtropical shallow-water fauna. Despite this, two species of *Conus*, *C. ebraeus* and *C. tesselatus*, which are Indo-Pacific in distribution, have been collected in East Pacific waters, whereas no species from the Caribbean are considered conspecific with any of the East Pacific *Conus*.

A few of the *Conus* species found in East Pacific waters seem more closely related to Atlantic species than to Indo-Pacific species. *Conus purpurascens* has a radular tooth structure which is virtually identical to that of *C. testudinarius* of the Atlantic, and both differ from the radular tooth structure of the Indo-Pacific fish-eating species such as *C. striatus, C. catus, C. consors, C. achatinus,* and *C. magnus,* all of which are virtually identical also. If it can be assumed that the radula is a conservative character which reflects relationships, then this is evidence of a close relationship of these two species. Kohn (1968) has noted that *C. perplexus* of the East Pacific and *C. jaspideus* of the Atlantic are virtually indistinguishable on the basis of shells, and Warmke's (1960) illustration of the radular tooth of *C. jaspideus* is similar to my own (Nybakken, 1970a) of *Conus perplexus*, which suggests another close relationship.

The shell of *C. austini* resembles closely that of *C. arcuatus*, and *C. re-

curvus has a shell which is closely similar to those of *C. cleri* and *C. villepini* of the Atlantic. All of these species have been collected by dredging. However, I have not seen radulas of the Atlantic forms, so it is speculation whether they can really be considered closely related.

Conus mus resembles *C. gladiator* both in shape and pattern of the shell and in radular structure. It seems likely that they are closely related. Similarly, *C. brunneus* has a radula virtually identical to that of *C. regius* of the Atlantic. The shape of the shell is also similar, though the pattern is somewhat different. The teeth of *C. brunneus* are also similar to those of *C. imperialis* and *C. zonatus* of the Indo-Pacific. In this case, however, the shape and pattern of the shell of *C. brunneus* are more similar to those of *C. regius* than to those of either *C. imperialis* or *C. zonatus* (Nybakken, 1970a).

Conus spurius of the Atlantic has a shell which resembles that of *C. regularis*, but the radulas of these two are quite different. The teeth of *C. spurius* (Warmke, 1960) resemble more closely those of *C. fergusoni*, which has a dissimilar shell.

In contrast to the above species, *C. tiaratus* seems most closely related to *C. miliaris* of the Indo-Pacific, and indeed it is difficult to tell the two apart on the basis of shells. The radulas are, however, slightly different.

I can say little further about the remaining species in both areas, due to lack of material.

It would seem on the basis of shell and tooth structure that at least a few species seem more closely related to Atlantic species than to Indo-Pacific species. However, I have not studied many of the Indo-Pacific species, so they may yet show resemblances which will indicate even closer relationships. Certainly the evidence that two Indo-Pacific species have crossed the East Pacific barrier to American shores suggests that others may also have done so in the past and may have evolved into forms which are now considered separate species. It is also possible that the radula is not a conservative character and should not be used, as here, to indicate relationships.

Sumario

Los Conidae de la Expedición del Pillsbury al Golfo de Panamá

Se obtuvieron 449 ejemplares de 11 especies de moluscos del género *Conus* en la expedición del barco de investigaciones John Elliott Pillsbury al Golfo de Panamá en 1967. De éstos, 446 ejemplares y diez especies fueron obtenidas dragando y rastreando en profundidades entre 15 y 210 m. Se revisa la literatura sobre conos que habitan por debajo de la marea, se discuten las diez especies y en algunos casos se amplían las descripciones y se ilustra *Conus poormani* Berry por primera vez. También se dan ilustraciones de las diez especies y se discuten las relaciones faunísticas de *Conus* en ambos lados del Istmo de Panamá.

LITERATURE CITED

BARTSCH, PAUL AND HARALD A. REHDER
    1939. Mollusks collected on the presidential cruise of 1938. Smithson. misc. Collns, 98(10): 1-18.

BERRY, S. STILLMAN
    1968. Notices of new eastern Pacific Mollusca. VIII. Leaflets in Malacology, 1(25): 155-158.

DALL, WILLIAM HEALEY
    1910. Summary of the shells of the genus *Conus* from the Pacific coast of America in the U. S. National Museum. Proc. U. S. natn. Mus., 38(1741): 217-228.

EKMAN, SVEN
    1953. Zoogeography of the sea. Sidgwick and Jackson, London, 417 pp.

EMERSON, WILLIAM K. AND L. G. HERTLEIN
    1964. Invertebrate megafossils of the Belvedere Expedition to the Gulf of California. Trans. S Diego Soc. nat. Hist., 13(17): 333-368.

EMERSON, WILLIAM K. AND WILLIAM E. OLD, JR.
    1962. Results of the Puritan–American Museum of Natural History Expedition to Western Mexico. 16. The Recent Mollusks: Gastropoda, Conidae. Am. Mus. Novit., No. 2112, 44 pp.

EMERSON, WILLIAM K. AND E. L. PUFFER
    1957. Recent mollusks of the 1940 "E. W. Scripps" cruise to the Gulf of California. Am. Mus. Novit., No. 1825, 57 pp.

HANNA, G. DALLAS
    1963. West American mollusks of the genus *Conus*. II. Occ. Pap. Calif. Acad. Sci., No. 35, 103 pp.

HANNA, G. DALLAS AND A. M. STRONG
    1949. West American mollusks of the genus *Conus*. Proc. Calif. Acad. Sci., Ser. 4, 26(9): 247-322.

HERTLEIN, L. G. AND A. M. STRONG
    1955. Marine mollusks collected by the "Askoy" Expedition to Panama, Colombia, and Ecuador in 1941. Bull. Am. Mus. nat. Hist., 107 (2): 165-317.

KEEN, A. MYRA
    1958. Seashells of tropical West America. Stanford University Press, Stanford, California, 624 pp.

KOHN, ALAN J.
    1959. The ecology of *Conus* in Hawaii. Ecol. Monogr., No. 29: 47-90.
    1968. Microhabitats, abundance and food of *Conus* on atoll reefs in the Maldive and Chagos Islands. Ecology, 49(6): 1046-1062.

NYBAKKEN, JAMES
    1970a. Correlation of radula tooth structure and food habits of three vermivorous species of *Conus*. Veliger, 12(3): 316-318.
    1970b. Radula anatomy and systematics of the West American Conidae (Mollusca, Gastropoda). Am. Mus. Novit., No. 2414, 29 pp.

PARKER, ROBERT H.
    1964. Zoogeography and ecology of macro-invertebrates of the Gulf of California and Continental Slope of western Mexico. *In*: Marine geology of the Gulf of California, a symposium. Mem. Am. Ass. Petrol. Geol., No. 3: 331-337.

ROBERTSON, ROBERT
1954. Suggested subgeneric allocations of Recent West American Conidae. Rep. Am. malac. Un., for 1954: 24.

STRONG, A. M. AND L. G. HERTLEIN
1939. Marine mollusks from Panama collected by the Allan Hancock Expedition to the Galapagos Islands, 1931-1932. Allan Hancock Pacif. Exped., 2(12): 177-245.

VON MOL, J. J., B. TURSCH, AND M. KEMPF
1967. Campagne de la Calypso au large des côtes Atlantiques de l'Amérique du Sud (1961-1962). 16. Mollusques prosobranches: les Conidae du Brésil. Ann. Inst. Oceanogr., New Series, 45(2): 233-254.

VOSS, GILBERT L.
1967. Narrative of R/V JOHN ELLIOTT PILLSBURY Cruise P-6703 in the Gulf of Panamá, April 29-May 11, 1967. Institute of Marine Sciences, University of Miami, 30 pp., 1 map. (Processed report.)

WAGNER, R. J. L. AND R. T. ABBOTT
1964. Van Nostrand's standard catalog of shells. Van Nostrand-Reinhold Books, Princeton, New Jersey, ix + 190 pp.

WARMKE, GERMAINE L.
1960. Seven Puerto Rico cones: notes and radulae. Nautilus, 73(4): 119-124.

BIOLOGICAL RESULTS OF THE UNIVERSITY OF MIAMI DEEP-SEA EXPEDITIONS. 79.

NEW AND UNUSUAL MOLLUSKS COLLECTED BY R/V JOHN ELLIOTT PILLSBURY AND R/V GERDA IN THE TROPICAL WESTERN ATLANTIC[1]

FREDERICK M. BAYER
Rosenstiel School of Marine and Atmospheric Sciences, University of Miami

ABSTRACT

Fifty-nine species of new or rare marine mollusks from the Caribbean area, 55 gastropods and four pelecypods, are reported and illustrated. Among these, the following new taxa are described: *Calliostoma olssoni*, n. sp., *Thelyssa callisto*, n. gen., n. sp., *Lischkeia deichmannae*, n. sp. (Trochidae); *Sconsia nephele*, n. sp. (Cassididae); *Typhis* (*Siphonochelus*) *tityrus*, n. sp. (Muricidae); *Columbarium* (*Peristarium*) *electra*, n. subgen., n. sp., *C.* (*P.*) *merope*, n. sp., *C.* (*P.*) *aurora*, n. sp. (Columbariidae); *Coralliophila fax*, n. sp., *C. sentix*, n. sp. (Coralliophilidae); *Teramachia chaunax*, n. sp. (Turbinellidae); *Lyria* (*Cordilyria*) *cordis*, n. subgen., n. sp., *Scaphella* (*Clenchina*) *evelina*, n. sp., *Volutomitra persephone*, n. sp., *V. erebus*, n. sp. (Volutidae); *Dimya tigrina*, n. sp., *Basiliomya goreaui*, n. gen., n. sp. (Dimyidae). The first record of living specimens of *Bathygalea coronadoi* (Crosse) is reported. *Mesorhytis meekiana* Dall, originally assigned to the family Fasciolariidae, is transferred to the genus *Teramachia* and placed in the family Turbinellidae. Shells are illustrated by photographs; radulae and opercula of some of the species are illustrated by drawings; and the gross anatomy of the mantle cavity is illustrated for *Oocorys sulcata* Fischer, *Vasum capitellum* (Linné), and *Lyria cordis*, n. sp.

INTRODUCTION

The faunal survey of the tropical West Atlantic, commenced in 1962 aboard R/V GERDA and in 1964 aboard R/V JOHN ELLIOTT PILLSBURY, has amassed an enormous amount of information concerning the distribution of invertebrates in this region. Because of the systematic diversity of the collections obtained, detailed studies of the entire fauna will be completed only over a long period of time. Many of the records, however, are of sufficient importance to warrant investigation in advance of full faunistic treatment. Some of these have revealed systematic problems of such magnitude that they cannot be dealt with adequately out of scientific

---

[1] Contribution No. 1311 from the University of Miami, Rosenstiel School of Marine and Atmospheric Sciences. This paper is one of a series resulting from the National Geographic Society–University of Miami Deep-Sea Biology Program. Operations of R/V PILLSBURY and R/V GERDA were supported by grants from the National Science Foundation: G-25376, GP-1363, GB-2204, GB-3808, GB-5776, GB-7082, GA-4569. Research was supported by a grant from the National Geographic Society for the study of deep-sea biology. Cost of the color figure was borne by the author.

context. Others have proved to be sufficiently clear-cut to permit treatment in preliminary fashion. Thus, it is the purpose of this paper to report new distributional records of certain rare or poorly known species and to describe some of the more important novelties that have come to light during processing of the collections.

Acknowledgments

This research has been supported by a grant from the National Geographic Society to the Rosenstiel School of Marine and Atmospheric Sciences, University of Miami, for investigation of the biology of the deep sea, Gilbert L. Voss, Principal Investigator. Support of shipboard operations was through NSF grants G-25376, GP-1363, GB-2204, GB-3808, GB-5776, GB-7082, and GA-4569. I am indebted to Dr. Voss for inclusion in this program, and to my colleagues at the Institute of Marine and Atmospheric Sciences for advice and assistance both tangible and intangible. The wholehearted cooperation of the captains and crews of R/V GERDA and R/V PILLSBURY has, from the outset, made shipboard exploration both enjoyable and successful.

It is a pleasure to extend a special word of appreciation to Axel A. Olsson, whose knowledge of the molluscan fauna of Middle America and its geological background has been an invaluable source of information, and whose association throughout this study has been a constant source of encouragement. Mrs. Jo Anne H. Romfh, Talbot Murray, and Dennis Opresko have rendered assistance in ways too numerous to mention. Mr. Robert C. Work has never failed to share his considerable knowledge of tropical American mollusks. Mrs. Constance Stolen McSweeny executed the drawings of opercula reproduced in Figures 10, 17, 22, 56, and 61.

Specimens collected by R/V OREGON along the coast of South America were obtained during EQUALANT cruises in which Donald R. Moore and Robert C. Work, of the School of Marine and Atmospheric Sciences, were members of the scientific party.

Systematic Part

The classification employed herein is basically that of Thiele (1929-35), incorporating modifications used by Keen (1958) and other current authors. It is by no means to be taken as authoritative, as classification of the higher taxa is not the aim of this paper. In general, however, I am unable to subscribe to the currently fashionable practice of generic splitting, as this commonly has been based upon insufficient biological grounds and thus serves no useful purpose.

The systematic disposition of the species included is as follows:

Class GASTROPODA
Family Trochidae
1. *Calliostoma (Kombologion) rosewateri* Clench & Turner
2. *Calliostoma (Kombologion) schroederi* Clench & Aguayo
3. *Calliostoma (s.l.) olssoni*, new species
4. *Calliostoma (Kombologion) hendersoni* Dall
5. *Lischkeia deichmannae*, new species
6. *Basilissa alta* Watson
7. *Thelyssa callisto*, new genus, new species
8. *Fluxina discula* Dall
Family Turbinidae
9. *Turbo (Halopsephus) haraldi* Robertson
Family Epitoniidae
10. *Epitonium (Amaea) mitchelli* (Dall)
11. *Epitonium (Solutiscala) vermetiforme* (Watson)
Family Cypraeidae
12. *Cypraea (Propustularia) surinamensis* Perry
Family Cassididae
13. *Bathygalea coronadoi* (Crosse)
14. *Sconsia striata* (Lamarck)
15. *Sconsia nephele*, new species
16. *Morum (Cancellomorum) dennisoni* (Reeve)
Family Oocorythidae
17. *Oocorys sulcata* Fischer
18. *Dalium solidum* Dall
Family Ficidae
19. *Ficus howelli* Clench & Aguayo
Family Muricidae
20. *Murex (Murex) olssoni* Vokes
21. *Murex (Murex) donmoorei* Bullis
22. *Murex (Murex) cabritii* Bernardi
23. *Murex (Siratus) beauii* Fischer & Bernardi
24. *Murex (Chicoreus) brevifrons* Lamarck
25. *Murex (Paziella) actinophorus* Dall
26. *Typhis (Siphonochelus) longicornis* Dall
27. *Typhis (Siphonochelus) bullisi* (Gertman)
28. *Typhis (Siphonochelus) tityrus*, new species
29. *Typhis (Talityphis) expansus* Sowerby
30. *Typhis (Pterotyphis) pinnatus* Broderip
Family Columbariidae
31. *Columbarium (Histricosceptrum) bartletti* Clench & Aguayo
32. *Columbarium (Fulgurofusus) bermudezi* Clench & Aguayo
33. *Columbarium (Fulgurofusus) brayi* Clench
34. *Columbarium (Peristarium) electra*, new subgenus, new species
35. *Columbarium (Peristarium) merope*, new species
36. *Columbarium (Peristarium) aurora*, new species
Family Coralliophilidae
37. *Coralliophila dalli* (Emerson & D'Attilio)
38. *Coralliophila mansfieldi* (McGinty)
39. *Coralliophila tectumsinensis* (Deshayes)
40. *Coralliophila sentix*, new species

41. *Coralliophila fax*, new species
42. *Coralliophila lamellosa* (Philippi)
43. *Coralliophila lactuca* Dall
    Family Buccinidae
44. *Phos beauii* Fischer & Bernardi
    Family Turbinellidae
45. *Teramachia meekiana* (Dall)
46. *Teramachia chaunax*, new species
47. *Turbinella laevigata* Anton
48. *Vasum capitellum* (Linnaeus)
    Family Volutidae
49. *Voluta virescens* Lightfoot
50. *Lyria* (*Cordilyria*) *cordis*, new subgenus, new species
51. *Scaphella* (*Scaphella*) *junonia* (Lamarck)
52. *Scaphella* (*Aurinia*) *dubia* (Broderip)
53. *Scaphella* (*Clenchina*) *evelina*, new species
54. *Volutomitra persephone*, new species
55. *Volutomitra erebus*, new species

Class PELECYPODA

Family Dimyidae
56. *Dimya argentea* Dall
57. *Dimya tigrina*, new species
58. *Basiliomya goreaui*, new genus, new species
    Family Spondylidae
59. *Spondylus gussoni* O. G. Costa

ZOOGEOGRAPHIC REMARKS

Previous work on the molluscan collections assembled in the course of faunal investigations at the Institute of Marine and Atmospheric Sciences has shown that the family Pleurotomariidae, long thought to be most extensively represented in the western Pacific from Japan southward and westward, is even better represented in the western Atlantic (Bayer, 1963, 1965, 1967). The recent discovery of a new species from Brazil (Moreira Leme & Penna, 1969) brings the total for this region to eight species, whereas six are known from Japanese and adjacent waters.

Another archaeogastropod genus associated chiefly with the Japanese fauna is *Lischkeia*, a new species of which has recently been trawled in the Antilles by R/V PILLSBURY. Among the turbinids, *Turbo haraldi* is remarkably similar to some well-known Indo-Pacific species.

Although species of the genus *Columbarium* are known from various parts of the world, *Columbarium pagoda* from Japan is perhaps the best known of all, so that reference to the genus immediately directs attention to the Japanese fauna. Four species with carinate shells have been taken in the western Atlantic by the BLAKE and the OREGON, and operations by the GERDA have now brought to light three species characterized by noncarinate shells. Thus, at least seven species of this genus occur in the

Caribbean region, eight if we accept the possibility that *C. sarissophorum* (Watson), originally from Pernambuco, also may occur there.

The genera *Coralliophila* and *Latiaxis* are also especially well represented in the Japanese fauna even though they are not by any means restricted to it. Nevertheless, the well-known *Coralliophila dalli* (Emerson & D'Attilio) is so similar to the Japanese *C. deburghiae* that even so experienced a zoologist as Dall misidentified it as that species. The discovery of two new species reported herein, also similar in many respects to Indo-Pacific species, serves to strengthen further the relationship of our fauna with that of the Pacific. It is to be expected that further exploration of the precipitous Antillean slopes abounding in corals will reveal additional species. Here, however, we are dealing with a group having a worldwide tropical distribution, and some western Atlantic species are closely related to, or identical with, eastern Atlantic and Mediterranean forms.

The presence of *Howellia*, now referred to the Indo-West Pacific genus *Teramachia*, in the Caribbean Sea has been known for thirty years (Clench & Aguayo, 1941), and its type-species, *H. mirabilis*, bears a strong resemblance to the Japanese *Teramachia tibiaeformis* Kuroda, the type-species of *Teramachia*, and to other large-shelled species from the Philippine Islands (*dalli, johnsoni, smithi*; see Weaver & du Pont, 1970). Specimens of *Mesorhytis meekiana* Dall and a related species taken by R/V PILLSBURY in the Caribbean show very close similarity of shell characters to the Philippine *Teramachia barthelowi* (Bartsch), and they accordingly are assigned to that genus. Their radular and opercular characters, now observed for the first time, indicate turbinellid rather than volutid affinities, but whether or not these small-shelled species of *Teramachia* will prove to be congeneric with the larger ones must remain for the future to reveal. Nevertheless, both groups are represented in both the Caribbean Sea and Indo-West Pacific waters.

Two new species of the genus *Volutomitra*, taken in rather deep water in the Caribbean Sea by R/V PILLSBURY, have their closest relationship with species from the Bering Sea and call to mind the distribution of the pelecypod *Thyasira disjuncta*, which occurs from Sitka, Alaska, to Coos Bay, Oregon, but found also by the PILLSBURY in the southwestern Caribbean (Boss, 1967).

Class GASTROPODA

Family Trochidae

1. *Calliostoma (Kombologion) rosewateri* Clench & Turner, 1960
Figs. 1, 2

Calliostoma (Kombologion) rosewateri Clench & Turner, 1960, Johnsonia, *4* (40): 41, pl. 6, fig. 3 (radula); pl. 10, fig. 2 (jaws); pl. 26 (shells).

FIGURE 1. Trochidae. *Calliostoma rosewateri* Clench & Turner: upper, from Sta. P-877, height 29 mm, diam. 32.5 mm; lower, from Sta. P-394, height 29.5 mm, diam. 34.4 mm.

FIGURE 2. Trochidae. *Calliostoma rosewateri* Clench & Turner; sculptural detail: left, from Sta. P-877; right, from Sta. P-394.

Material Examined.—PILLSBURY Sta. P-394. Caribbean Sea west of Cartagena, Colombia: 9°28.6'N, 76°26.3'W, 421-641 meters, 16 July 1966. One specimen, height 29.5 mm, greatest diameter 34.4 mm.—PILLSBURY Sta. P-877. Lesser Antilles, ESE of Bequia Island: 13°16.7'N, 61°05.6'W, 329-467 meters, 6 July 1969. One specimen, height 29 mm, maximal diameter 32.5 mm.

Remarks.—Two living specimens about the same size as the type-material were obtained by R/V PILLSBURY. One of these, from off Cartagena, resembles the type in all respects (Figs. 1 [lower], 2 [right]). The other, from the Lesser Antilles, differs slightly as to details. The two beaded cords separating the three strong ones above the periphery are absent and the distinct subsutural plain cord is poorly developed (Fig. 2 [left]). The color pattern on the spire is identical in the two specimens, but that from the Lesser Antilles has only four, very distinct, reddish brown spiral lines on the base (Fig. 1 [upper]) compared with 5 or 6 in the types and 7 in the example from station P-394, in which those toward the periphery are broader and more diffuse (Fig. 1 [lower]).

These two records extend the geographical range of the species to the westward and to the northward, and demonstrate somewhat more variation than was indicated by the type-material.

2. *Calliostoma (Kombologion) schroederi* Clench & Aguayo, 1938
Fig. 3

Calliostoma schroederi Clench & Aguayo, 1938, Mem. Soc. Cubana Hist. nat., *12*: 377, pl. 28, fig. 3.—Clench & Turner, 1960, Johnsonia, *4*(40): 45, pl. 7, fig. 2 (radula); pl. 11, fig. 1 (jaws); pl. 29 (shells).

Material Examined.—GERDA Sta. 915. Bahama Islands, Northwest Providence Channel: 25°54'N, 78°12'W, 439 meters, 26 September 1967. One specimen, height 20 mm, greatest diameter 24 mm.

Remarks.—Except for a specimen taken by R/V COMBAT (Sta. 235) off Matanilla Shoal, Little Bahama Bank, this species is known from the Old Bahama Channel off the north coast of Cuba. The present record from the Bahamas suggests that the species is probably generally distributed through this area on the proper substrate between roughly 250 and 450 meters.

3. **Calliostoma** *(s.l.)* **olssoni**, n. sp.
Fig. 4, (left)

Description.—Shell trochoid, solid, umbilicate, rather strongly sculptured. Whorls eight, flat sided, distinctly keeled. Spire moderately extended, weakly concave, produced at an angle of 84°. Nuclear whorls about 1½, worn but apparently developing cancellate sculpture quite early. Base moderately

FIGURE 3. Trochidae. *Calliostoma schroederi* Clench & Aguayo, Sta. G-915, height 20 mm, diam. 24 mm.

convex. Aperture subquadrate, outer lip evidently simple (broken in the unique type). Columella weakly arched, anterior truncated. Body whorl with ten beaded spiral cords above the periphery, all but the lowest one white, the interspaces orange-brown; cord adjacent to periphery narrower, indistinctly beaded, colored like the area to either side. Peripheral cord strong but not beaded, canted somewhat apically; rounded basal part of peripheral keel with four or five indistinct raised spiral lines, remainder of base smooth and polished, marked with microscopic radial lines of growth. Spire with pinkish brown axial blotches that somewhat discolor the white spiral cords; periphery with a series of purplish pink blotches which extend inward on the base toward the umbilicus as curved or angular

FIGURE 4. Trochidae. Left, *Calliostoma olssoni*, n. sp., holotype, Sta. P-876, height 16.8 mm, diam. 21 mm; right, *Calliostoma hendersoni* Dall, Sta. T-10.

rays on a white ground color; base with five yellow-brown spiral stripes, the peripheral and umbilical ones the weakest. Umbilicus rather widely open, deep, bounded by a white spiral cord of smooth callus extending from the end of the columella.

Measurements.—Height 16.8 mm, greatest diameter 21 mm.

Holotype.—USNM No. 700002, PILLSBURY Sta. P-876.

Type-Locality.—PILLSBURY Sta. P-876. Lesser Antilles, SW of St. Vincent, 13°13.9′N, 61°04.7′W, 231-258 meters, 6 July 1969.

Remarks.—Similar to *Calliostoma hendersoni* Dall in general appearance, but differs by its smooth base, more numerous spiral cords, and striking coloration.

Known only from a single, somewhat damaged specimen.

4. Calliostoma (Kombologion) hendersoni Dall
Fig. 4, (right)

Calliostoma hendersoni Dall, 1927, Proc. U. S. natn. Mus., *70*: 7.
Calliostoma (Kombologion) hendersoni, Clench & Turner, 1960, Johnsonia, 4(40): 43, pl. 7, fig. 4; pl. 11, fig. 2; pl. 28.

Records.—From five stations of R/V GERDA and one of R/V TURSIOPS: G-134, Straits of Florida off the lower keys (position not recorded, 191 m, 21 June 1963; 1).—G-482, Straits of Florida (24°29'N, 80°54'W, 201-210 m, 26 January 1965; 1).—G-813, Straits of Florida (24°31.5'N, 80°40'W, 201 m, 21 June 1967; 3).—G-839, Straits of Florida (24°23'N, 80°52'W, 236-229 m, 11 July 1967; 1).—G-866, Straits of Florida (24° 28'N, 81°09'W, 186 m, 29 August 1967; 1).—T-10, off Alligator Reef Lighthouse (position not recorded, 133-154 m, 23 June 1966; 1).

Remarks.—Previously known only from two localities in the Straits of Florida (8 miles SE of Key West in 118 fms = 216 m; and 8 miles NE of Cay Sal Bank, 24°03'N, 80°30'W, 150 fms = 274 m, OREGON Sta. 1349).

5. Lischkeia deichmannae n. sp.
Fig. 5

Description.—Shell elevated, trochoid in shape, conspicuously sculptured, rather large but thin and light in construction; height 54 mm, maximal diameter 42 mm. Nuclear whorls missing, about 8½ postnuclear whorls remaining. Spire extended, produced at an angle of 58°; suture impressed in a distinct channel; whorls convex, shouldered. Two spiral cords set with prominent conical nodules delimit a broad, flat periphery. A pair of weak spirals, thickened at intervals where they are crossed by axial lamellae, bounds a narrow, flat shelf adjacent to the suture, which is developed a short distance below the lower peripheral cord. This results in a distinct channel containing the suture, which develops along a spiral that becomes nodose on the last half of the body whorl and forms a distinct boundary for the base. Base moderately convex, ornamented with ten spiral cords that become nodose toward the edge of the outer lip. Axial sculpture consists of conspicuous, thin, undulating and anastomosing, raised lamellae which, in turn, are marked with microscopic growth lines. Aperture subcircular, with a thickened, reflected outer lip evidently indicative of maximal growth, produced at an angle of about 45° from the base. Columella arched, smooth, without teeth. Parietal wall with a thin glaze that is nacreous on the pillar and in an area adjacent to the sutural angle, but porcellaneous in the intervening area. Inside of outer lip nacreous. Outer surface creamy white, semiopaque, possibly allowing a nacreous lustre to show through in living material.

Holotype.—USNM No. 700003, Sta. P-889.

FIGURE 5. Trochidae. *Lischkeia deichmannae*, n. sp., holotype, Sta. P-889, height 54 mm, diam. 42 mm.

Type-Locality.—R/V PILLSBURY Sta. P-889, off St. Lucia, Lesser Antilles: 14°04.4'N, 60°50.8'W, 371-403 meters, 7 July 1969.

Record.—GERDA Sta. G-897, off Arrowsmith Bank, Yucatan, 20°59'N, 86°24'W, 293-210 m, 10 September 1967. A fragmentary specimen consisting of the thickened palatal lip and part of the base.

Comparisons.—On first sight, *Lischkeia deichmannae* is strikingly similar to the type-species, *L. monilifera* (Lamarck) (= *L. alwinae* Lischke) from Japan. Comparison of specimens reveals significant differences, however. In *L. deichmannae* the spire is more extended, the base more inflated and convex, and the umbilical callus completely seals over the umbilical depression. In *L. monilifera*, the edge of the callus is free so that the shallow spiral channel bordering the columella is roofed over by its advancing edge but remains open terminally. Also, in *L. monilifera* the spiral cords are more distinct, beaded at their intersections with narrow axial cords. In *L. deichmannae*, the spiral sculpture is obscure except for the two prominently nodulose cords above and below the periphery and those on the base, and the lamellate axial sculpture is very conspicuous. Specimens of *L. monilifera* from Sagami Bay, Honshu, Japan, in the collections of the

National Museum of Natural History (343262) have a height of 37.6 mm, diameter 33.8 mm, angle of spire 73°. From Tosa Bay, Shikoku, 70 fms = 128 m (605948), height 45.5 mm, diameter 44 mm; height 36.7 mm, diameter 32.8 mm; and height 37.8 mm, diameter 34.5 mm. From Tanabe, Kii, Honshu (273648), height 35.2 mm, diameter 31.5 mm; and height 35.8 mm, diameter 32 mm.

Lischkeia deichmannae more closely resembles a specimen of *Lischkeia* from off Ternate (ALBATROSS Sta. 5617, 131 fms = 240 m; USNM No. 239250), which has a similarly elevated spire, more convex base, and raised axial lamellae. In this example, however, the spiral sculpture is distinctly developed, the base is not so sharply set off from the side of the body whorl, and the major spiral cords on the peripheral half of the base are strongly nodose.

Genus *Basilissa* Watson

Basilissa Watson, 1879, Journ. Linn. Soc. Lond. *15*: 593; 1886, Rep. Sci. Res. Challenger, Zool., *15* (Part 42): 96.—Dall, 1889a, Bull. Mus. comp. Zool. Harv., *18*: 383.—Cotton, 1959, South Australian Mollusca, Archeogastropoda: 189.—Knight et al., 1960, Treat. Invert. Paleont. *1*(1): 250.

Type-Species.—*Basilissa superba* Watson, by subsequent designation: Cossmann, 1888.

Description.—See Dall, 1889a; Cotton, 1959.

Remarks.—These shells resemble a small, iridescent *Calliostoma*, but are characterized by a wide, moderately deep sinus in the outer lip near the suture and another below the periphery, in the basal part of the margin, resulting in a clawlike peripheral projection of the outer lip.

Infrequently noticed because of their small size, specimens of these shells have been secured by R/V PILLSBURY and R/V GERDA at a number of stations in the Straits of Florida. Members of the related genera *Seguenzia* and *Fluxina* also were obtained. As in previous collections, most of the specimens were dead and somewhat damaged, although a few examples of *Basilissa* were taken alive. The material shows that *Fluxina discula* Dall differs in no significant regard from the genus *Basilissa*. The shells are distinctly nacreous when wet, although they take on a porcellaneous appearance upon drying. Moreover, *Seguenzia costulata* differs from *Basilissa* only in having a stronger columellar fold and more deeply sinuate lip, thus forming a transition between the genera as already noticed by Dall.

6. *Basilissa alta* Watson
Figs. 6, D-G; 7

Basilissa alta Watson, 1879, Journ. Linn. Soc. Lond., *15*: 597.—Dall, 1881, Bull. Mus. comp. Zool. Harv., *9*(2): 48; 1889a, Bull. Mus. comp. Zool. Harv., *18*: 384.

Seguenzia delicatula Dall, 1881, Bull. Mus. comp. Zool. Harv., *9*(2): 48.
Basilissa alta var. *Oxytoma* Watson, 1885, Rep. Sci. Res. Challenger, Zool. *15* (part 42): 100, pl. 7, fig. 8e.
Basilissa alta var. *delicatula* Dall, 1889a, Bull. Mus. comp. Zool. Harv., *18:* 384, pl. 22, figs. 2, 2a; 1889b, Bull. U. S. natn. Mus. *37*: 164, pl. 22, figs. 2, 2a (no descr., fig. from 1889a).

Material Examined.—Twenty-seven specimens from the following stations of R/V PILLSBURY and R/V GERDA: P-861 (12°42′N, 61°05.5′W, 18-744 m, 4 July 1969; 1).—P-1255 (17°18′N, 78°32′W, 805-722 m, 14 July 1970; 2).—P-1261 (17°13′N, 77°50′W, 722-768 m, 15 July 1970; 4).—G-365 (24°11′N, 81°37′W, 672 m, 15 September 1964; 1).—G-370 (23°54′N, 81°19′W, 1281 m, 16 September 1964; 1).—G-478 (24°15′N, 82°11′W, 543-348 m, 26 January 1965; 1).—G-815 (24°08′N, 79°48′W, 618 m, 22 June 1967; 1).—G-959 (23°25′N, 82°35′W, 1829 m, 31 January 1968; 5).—G-960 (23°30′N, 82°26′W, 1697-1692 m, 31 January 1968; 6).—G-964 (23°46′N, 81°51′W, 1390-1414 m, 1 February 1968; 1).—G-966 (24°10′N, 82°22′W, 553-558 m, 2 February 1968; 1).—G-967 (24°15′N, 82°26′W, 499-503 m, 2 February 1968; 3).

Remarks.—All of the specimens noted above agree well with the descriptions given by Watson (1878, 1885) for the species and by Dall (1881, 1889a) for the "variety" *delicatula*. The latter was distinguished from *alta* (Dall, 1889a: 384) by its thinner shell and the presence of fine spirals over the whole surface of the whorls. As Watson recognized a variety *oxytoma* with sculpture more distinct than in *alta*, and as the present material shows considerable variation in this regard, it is preferable to treat them all under the name *alta*. The specimens vary also as to relative height of the shell. In Watson's original material from off Culebra Island, east of Puerto Rico, $^h/_d = 1.04$, whereas in Dall's *delicatula* it is 0.83. In the present material, $^h/_d$ ranges from 0.82 (Sta. G-964) to 0.99 (Sta. G-966), very closely paralleling the specimens previously reported.

The radula (Fig. 7) has a rachidian with a triangular cusp finely denticulated on the sides, a wide lateral with an inwardly directly triangular cusp denticulated on both sides, and several (6 or 7) marginals, flat and rather narrow, denticulated along most of the outer edge but on the inner edge only near the tip.

The operculum is circular, very thin, concave, of about four whorls.

Thelyssa, new genus

Description.—Shell resembling that of *Basilissa*, nacreous under a porcellaneous layer, small, conical, with nearly flat base and rhomboidal aperture, broadly umbilicate; outer lip with a shallow sinus adjacent to the suture and another, broad and shallow, in the peripheral half of the basal part;

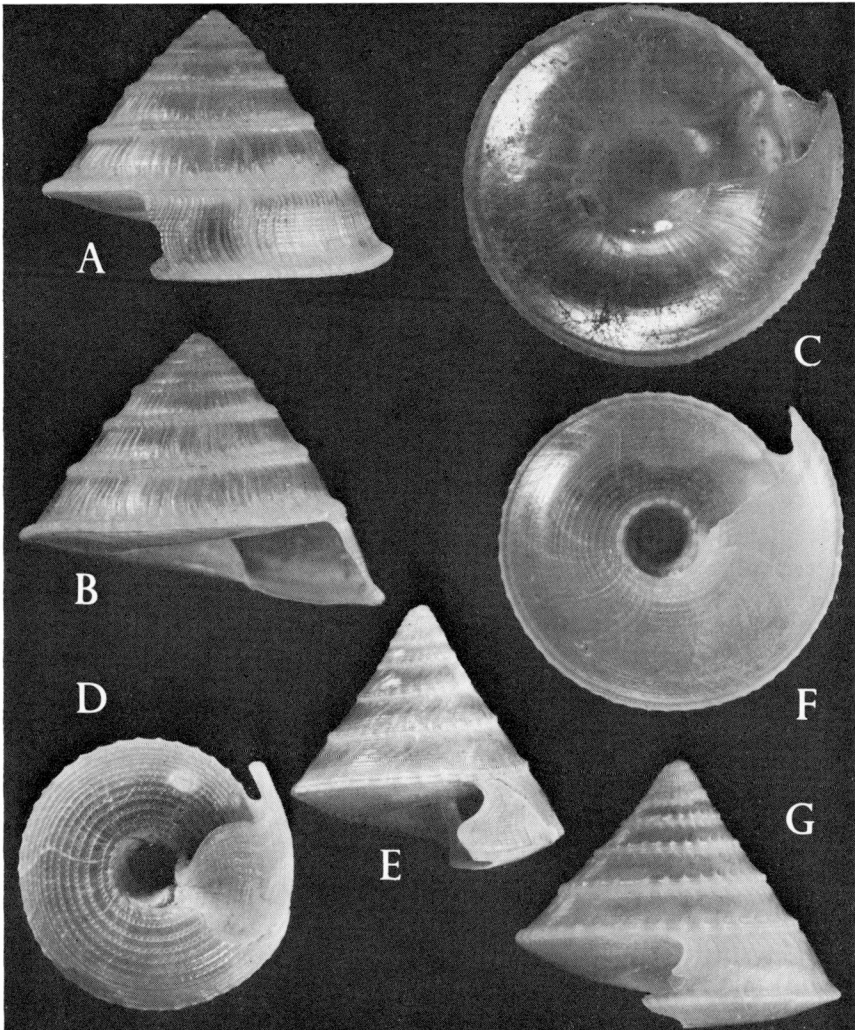

FIGURE 6. Trochidae. A-C, *Thelyssa callisto*, holotype, Sta. P-1138, height 5.8 mm, diam. 7.6 mm.—D-E, *Basilissa alta* Watson, Sta. P-861, height 5.55 mm, diam. 5.8 mm.—F-G, *Basilissa alta* Watson ("var. *delicatula*" Dall), Sta. G-964, height 5.75 mm, diam. 6.95 mm.

columella with a blunt terminal tubercle but not strongly toothed, columellar lip reflected as a band of callus within the umbilicus, subsequently forming a callous septum completely sealing the umbilical opening in fully developed shells.

Type-Species.—*Thelyssa callisto,* n. sp., here designated.

Gender.—Feminine.

Remarks.—The peculiar band of callus laid down within the umbilicus behind the columella, and the final closure of the umbilicus by a smooth callous septum are highly distinctive features; the former extends high within the spire and would be visible even in young shells without the umbilical septum or in shells having the septum broken away.

7. **Thelyssa callisto**, new species
Fig. 6, A-C

Description.—Shell small, conical, spire with weakly convex sides, base almost flat; glossy, translucent, white, brilliantly iridescent when wet but outer layer becoming milky and porcellaneous when dry, thus obscuring the iridescence. Whorls eight: nucleus of about 1¼ smooth, rounded, glassy whorls; postnuclear whorls sculptured, peripherally carinate, weakly concave above the carina and weakly convex below the suture. Postnuclear whorls with numerous narrow, sigmoid axial ribs corresponding in contour with the microscopic lines of growth, each forming a low, rounded, oblique ridge where it passes over the periphery, continuing indistinctly across the base in a wide sigmoid curve. First five postnuclear whorls without spiral sculpture except for three narrow raised cords on the carina; late in the seventh whorl, low, narrow spiral threads appear above the peripheral carina, increasing in number to about 14 immediately behind the outer lip. Base with exceedingly fine spiral threads, those near the periphery more distinct than the rest. Umbilicus funnel-shaped, wide, bounded by a strong spiral cord beaded by the axial (i.e., transverse) riblets; within the umbilicus, sutures between the whorls deeply impressed. Aperture rhomboidal; outer lip sharp, with a shallow sinus adjacent to the suture and a broader one on the base near the periphery; columellar lip oblique, terminating in a weak, blunt denticulation, reflected inward as a tongue of callus forming a spiral band within the umbilicus, in the adult shell producing also a smooth, solid callous pad that closes the umbilicus completely; parietal wall with a thick glaze of transparent callus.

Measurements.—Height 5.8 mm, diameter 7.6 mm (holotype); height 5.5 mm, diameter 7.8 mm (paratype); height 5.8 mm (apex damaged), diameter 7.3 mm (paratype).

Holotype.—USNM No. 701215, PILLSBURY Sta. P-1138.

Type-Locality.—PILLSBURY Sta. P-1138, W of Great Inagua Island, 20°51.7'N, 74°22'W, 2745-2751 meters, 12 January 1970.

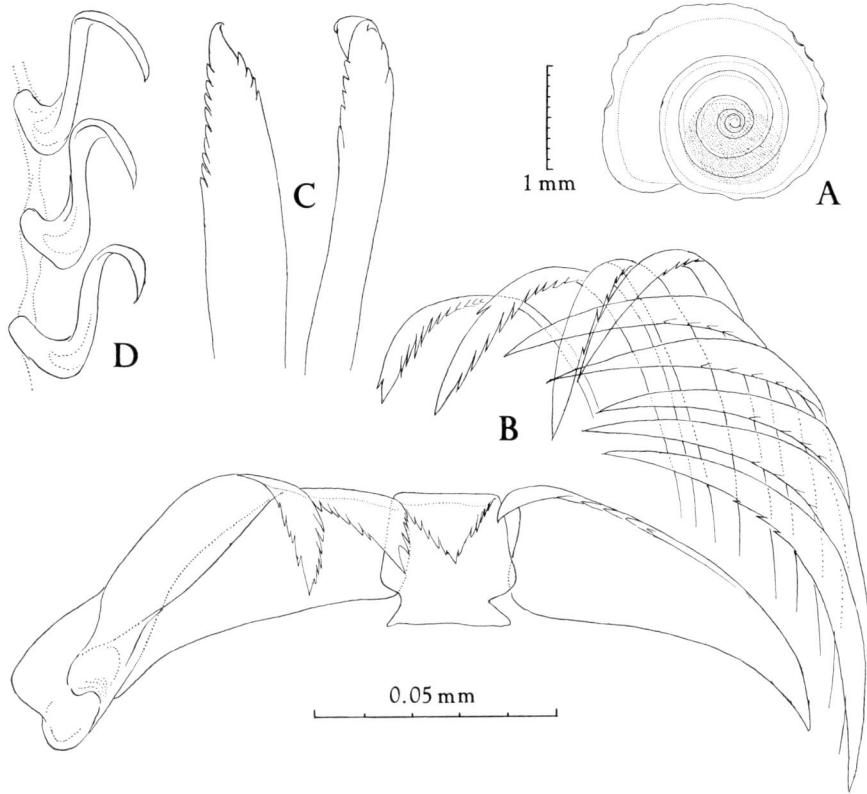

FIGURE 7. Trochidae. *Basilissa alta* Watson, Sta. P-1255: A, operculum; B, radula, bases of marginals not visible (scale applies to all radulae); C, tips of two marginals; D, profile view of rachidians.

Remarks.—This species resembles *Basilissa alta* Watson in size and shape, but the sides of the spire are faintly convex instead of flat or weakly concave, the sigmoid axial riblets are more pronounced, the base is only faintly sculptured, the position of the sinuses differs with the result that the most advanced part of the outer lip is above the periphery, and, above all, the umbilicus is closed by a smooth callous plate in the fully developed shell.

Genus *Fluxina* Dall

Fluxina Dall, 1881, Bull. Mus. comp. Zool. Harv., 9(2): 51.

Description.—Dall, 1881.

Type-Species.—*Fluxina brunnea* Dall, by monotypy.

FIGURE 8. Trochidae. *Fluxina discula* Dall, Sta. G-967, height 2.7 mm, diam. 6.4 mm.

Remarks.—The type-species has a large, *Calliostoma*-like shell. *Fluxina discula* Dall, 1889, obviously is close to *Basilissa* Watson, differing chiefly in its depressed spire. As not enough material is at my disposal to determine whether the two species really are congeneric, *F. discula* is provisionally retained in *Fluxina.* It is considered to be trochid, not architectonicid, in its affinity.

8. *Fluxina discula* Dall
Fig. 8

Fluxina discula Dall, 1889a, Bull. Mus. comp. Zool. Harv., *18*: 273, pl. 23, figs. 5-6; 1889b, Bull. U. S. natn. Mus., *37*: 148, pl. 23, figs. 5-6 (listed only; figure from 1889a).

Material Examined.—GERDA Sta. G-967, Straits of Florida SW of Marquesas Keys, 24°15'N, 82°26'W, 499-503 m, 2 February 1968; 1 specimen.

Remarks.—Nacreous iridescence clearly shows through the white outer layer of shell when wet, but when dry the shell is opaque, glossy white. Although this specimen was not alive when collected and therefore cannot provide radular characters which would assist in systematic placement, its conchological characters are so distinctly like *Basilissa* that only the great depression of the spire justifies maintaining it apart from that genus.

Fluxina discula was obtained originally from the vicinity of Dominica.

Family Turbinidae
9. *Turbo* (*Halopsephus*) *haraldi* Robertson
Figs. 9, 10

Halopsephus pulcher Rehder, 1943, Proc. U. S. natn. Mus., *93*(3161): 191, pl. 20, figs. 3, 10.
Turbo (*Halopsephus*) *haraldi* Robertson, 1957, J. Wash. Acad. Sci., *47*(9): 316, figs. 1-3.
Not *Turbo pulcher* Dillwyn, 1817, Catalogue Shells, 2: 855.
Not *Turbo pulcher* Reeve, 1842, Conch. Systematica, 2: 167.

R/V PILLSBURY obtained this rare species at five stations in the Caribbean area, substantially increasing the known range and suggesting a general distribution in this area. Fifteen specimens, nine in good condition (two collected alive), show that individuals attain a height of 25 mm, over twice the size indicated in the two previous records of the species.

Records.—Sta. P-409, eastern Panama (8°51.2'N, 77°28.1'W, 54-47 m; one dead specimen, height 9 mm, width 8.6 mm).—P-420, off Archipielago de Mulatas, Panama (9°30.5'N, 78°25.6'W, 50 m; one live specimen, height 19.8 mm, width 18 mm).—P-857, east of Carriacou, Lesser Antilles (12°23.5'N, 61°21.6'W, 9-348 m; three dead specimens; height 22.2 mm,

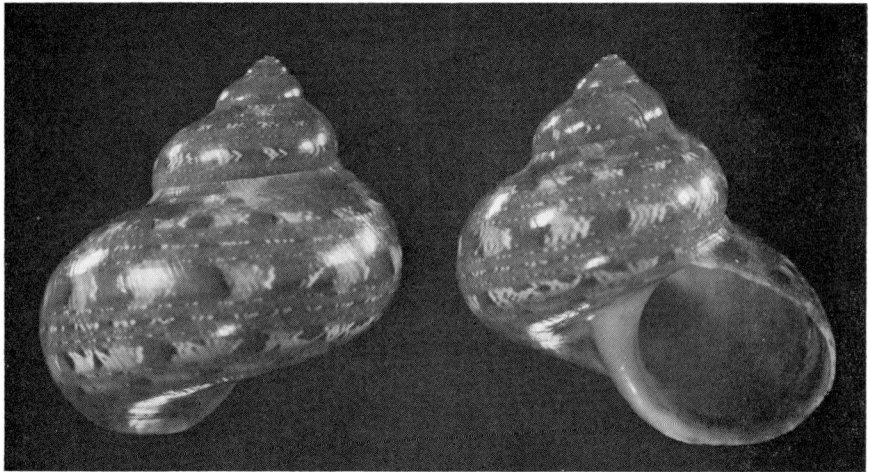

FIGURE 9. Turbinidae. *Turbo haraldi* Robertson, Sta. P-924, height 20.4 mm, width, 18.6 mm.

width 19.6 mm; height 5.9 mm, width 6.1 mm; height 5.4 mm, width 5.9 mm).—P-924, Dominica Channel, Lesser Antilles (15°13.0'N, 60°56.9'W, 68-69 m; one live specimen, height 20.4 mm, width 18.6 mm; one dead specimen, height 19.9 mm, width 18.6 mm).—P-926, Dominica Channel, Lesser Antilles (15°13.2'N, 60°56.8'W, 73 m; eight dead specimens, ranging in size from 25 mm high, 22.5 mm wide, to 18.5 mm high, 16.8 mm wide).

Remarks.—The present material shows that adult shells have not been reported previously. The shells strongly resemble small specimens of *Turbo petholatus* Linné and *Turbo reevei* Philippi, but are more reddish or orange in color. The basic color pattern consists of three spiral bands of color darker than the ground color, one on the obscure subsutural spiral cord, one at the periphery, and one equally spaced below the periphery; in some specimens, the base is paler below the lowest spiral band. The spiral bands are interrupted by regularly spaced paler zones which may contain even paler crescentic lines as wide as the band, or groups of pale vertical flecks. Three narrower, less prominent dark spirals separate the adjacent major spirals, and even weaker ones may occur between these; all are interrupted by pale flecks. Narrow, pale axial lines run in the direction of the growth lines and, in some specimens, there are narrow, curved, pale marks extending down from the suture to the subsutural cord. Slight variations in background color produce a generally clouded effect; the color overall may be bright reddish, rusty or ochraceous yellow. Robertson (1957:318) dis-

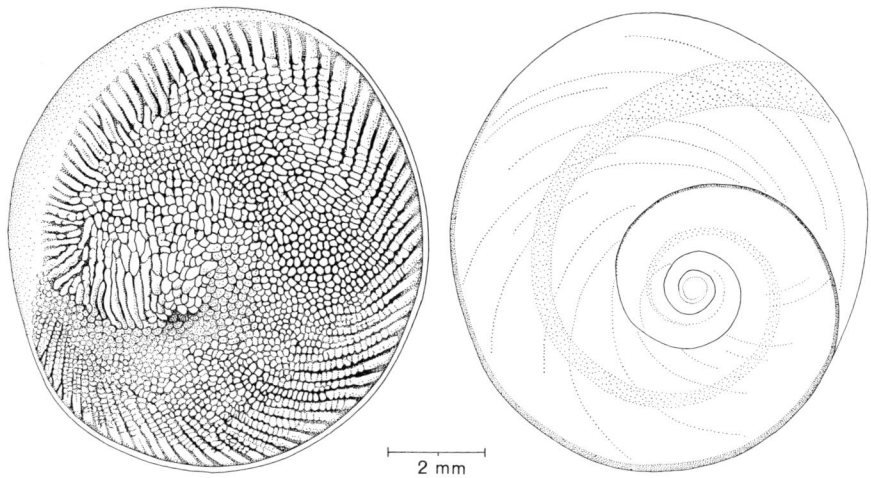

FIGURE 10. Turbinidae. *Turbo haraldi* Robertson, Sta. P-420; operculum. Drawing by Constance S. McSweeny.

tinguished juveniles of *T. haraldi* from those of *T. castanea* by their bright carmine color. However, his intention is not clear as he also described the early nuclear whorls as white. In the present material, the first nuclear whorl is pink, brownish orange, or white, the following whorls pink with more or less white clouding; in a specimen from station P-420, the nucleus and two postnuclear whorls are white, becoming pink in the third whorl.

In some examples, the narrow umbilicus is not closed by the columellar callus.

The distinctive adult operculum, which has not been illustrated previously, is shown in Figure 10. It provides the chief distinguishing character of the subgenus *Halopsephus*, which otherwise is like *Turbo s.s.*

Range.—Bahamas (Robertson), to the Lesser Antilles (Rehder) and Panama (here recorded).

Type-Locality.—Barbados.

Family Epitoniidae

10. *Epitonium (Amaea) mitchelli* (Dall)
Fig. 11, A

Scala mitchelli Dall, 1896, Nautilus, 9: 112.
Scala (Amaea) mitchelli, Dall, 1902, Proc. U. S. natn. Mus., 24(1264): 506, pl. 30, figs. 3-4.

FIGURE 11. Epitoniidae. A, *Epitonium mitchelli* (Dall), Sta. P-324, height 45.2 mm, diam. 16.1 mm; B, *Epitonium brunneopictum* Dall, Sta. P-515, Bay of Panama, height 35.1 mm, diam. 10.9 mm; C, *Epitonium ferminianum* Dall, Sta. P-532, Bay of Panama, height 56.2 mm, diam. 23.5 mm; D, *Epitonium vermetiforme* (Watson), Sta. G-966, height 8.75 mm.

Amaea (*Amaea*) *mitchelli*, Clench & Turner, 1950, Johnsonia, 2(29): 243, pl. 106, figs. 5-7 (photo of holotype, additional specimen, and sculptural detail).

Description.—Clench & Turner, 1950.

Holotype.—USNM No. 187792.

Type-Locality.—Matagorda Island, Texas.

Record.—PILLSBURY Sta. P-324. Caribbean coast of Panama: 9°44'N, 79°31'W, 64-55 m, 7 July 1966. One dead specimen, bored by predaceous gastropod but in fresh condition; height 45.2 mm, diameter of body whorl 16.1 mm.

Remarks.—So far as we are aware, this is the first record of this species in the southwestern Caribbean Sea.

Epitonium (*Amaea*) *mitchelli* bears a strong resemblance to the eastern Pacific *Epitonium* (*Ferminoscala*) *ferminianum* Dall, 1908, (Fig. 11, C) and, especially, *E.* (*Ferminoscala*) *brunneopictum* Dall, 1908, (Fig. 11, B) in size, shape, sculpture, and coloration. Although placed in different genera (or subgenera) by recent authors because of the absence of varices, examination of *E. ferminianum* and *E. brunneopictum* shows that in both species the axial lamellae are stronger at irregular intervals, almost as conspicuously as in *E. mitchelli*. Further, the basal area of the body whorl in *E. mitchelli* is set off by stronger spirals as it is in the two eastern Pacific species, although perhaps not so conspicuously. It appears that the more pronounced differentiation of the "basal disk" in *Scalina* (*Ferminoscala*) has been overemphasized in the search for differences. In my opinion, the distinction between *Amaea* H. & A. Adams, 1853, and *Scalina* Conrad, 1865, (= *Ferminoscala* Dall, 1908) is not justified.

11. *Epitonium* (*Solutiscala*) *vermetiforme* (Watson)
Fig. 11, D

Scalaria vermetiformis Watson, 1886, Rep. Sci. Res. Challenger, Zool., *15*: 142, pl. 9, fig. 6.
Solutiscala (*Solutiscala*) *vermetiformis*, Clench & Turner, 1952, Johnsonia, 2(31): 347, pl. 170, figs. 1-2 (original figures reproduced).

Description.—See Clench & Turner, 1952: 347.

Record.—GERDA Sta. G-966. Straits of Florida SW of Marquesas Keys, 24°10'N, 82°22'W, 544-549 meters, 2 February 1968. One dead but nearly perfect specimen, length 8.75 mm, width 3.5 mm.

Remarks.—This species has not been reported since its original description by Watson in 1866. The rather small size and very delicate structure of the shell no doubt account for the lack of subsequent reports.

A shell very similar to this was described as *Delphinula nitida* by Verrill & Smith from ALBATROSS Sta. 2229 (Verrill, 1885: 424, pl. 44, fig. 11). Dall (1889b, pl. 46, fig 11) reproduced Verrill's plate and referred the species to *Laxispira*. Although the spire of that species is not so extended as in *E. vermetiforme*, the shells are otherwise very close, and it seems likely that they are related if not identical.

Family Cypraeidae
12. *Cypraea* (*Propustularia*) *surinamensis* Perry
Fig. 12

Cypraea surinamensis Perry, 1811, Conchology: pl. 20, no. 4.—Burgess, 1970, Living Cowries: 237, pl. 23, fig. B.
Propustularia surinamensis, Coomans, 1963, Stud. Fauna Curaçao, 15: 63, pl. 2, e-f.
Cypraea (*Propustularia*) *surinamensis*, Emerson & Old, 1965, Nautilus, *79* (1): 26-30, pl. 3, figs. 1-2 (new records; nomenclatural discussion); 1966, Nautilus, *80*(2): 70-71 (species reported from Brazil).—Matthews, 1967, Arq. Est. Biol. Mar. Univ. Fed. Ceará, 7(1): 17, figs. 7-8.

FIGURE 12. Cypraeidae. *Cypraea surinamensis*, Sta. P-581, height 37 mm, width 21 mm.

Record.—PILLSBURY Sta. P-581, off Arrowsmith Bank, Yucatan, 21°05'N, 86°23'W, 146-265 m, 22 May 1967. One specimen, "long form," length 37 mm, width 21 mm.

Remarks.—The nomenclatural difficulties revolving around this species have been discussed by Coomans (1963) and Emerson & Old (1965).

Modern records from the Florida Keys, Yucatan, and Brazil confirm that *C. surinamensis* has a rather wide range in the tropical western Atlantic.

Family Cassididae

Genus *Bathygalea* Woodring & Olsson, 1957

Bathygalea Woodring & Olsson, 1957, Prof. Pap. U. S. Geol. Survey No. 314-B: 22.

Description.—See Woodring & Olsson, 1957.

Type-Species.—*Cassis coronadoi* Crosse, by original designation.

13. *Bathygalea coronadoi* (Crosse, 1867)
Figs. 13; 17, C

Cassis coronadoi Crosse, 1867, Journ. Conchyl., *15*: 64, pl. 4, fig. 1; pl. 5, fig. 1 (Matanzas, Cuba).
Galeodea coronadoi, Dall, 1889a, Bull. Mus. Comp. Zool. Harv., *18*: 231 (40 miles off Cape Fear, N. C., 124 fathoms, ALBATROSS Sta. 2603).— Clench, 1944, Johnsonia, *1*(16): 4, pl. 2 (citations of earlier records).
Bathygalea coronadoi, Woodring & Olsson, 1957, Prof. Pap. U. S. Geol. Survey No. 314-B: 24, pl. 9, figs. 2-3 (description and figure of specimen reported by Dall, 1899a).

Material Examined.—PILLSBURY Sta. P-739. Venezuela, NW of Isla Centinela: 10°54.7′N, 66°17.8′W, 234-280 meters, 23 July 1968. Two living specimens: length 62.9 mm, width 42.7 mm; length 49.5 mm, width 34.8 mm.

R/V JOHN ELLIOTT PILLSBURY obtained two living examples of *Bathygalea coronadoi* (Crosse) at trawling station P-739 about 50 miles northeast of Caracas, Venezuela, the third record of this species and the first specimens to be obtained alive. This record extends the known geographical range considerably to the southward and indicates a general distribution in the Caribbean area. Although Woodring & Olsson attributed the paucity of records to inadequate exploration in the tropical western Atlantic, the many trawling stations occupied in the region by vessels of the School of Marine and Atmospheric Sciences without obtaining additional specimens seem to indicate scarcity.

Because of the great rarity of *Bathygalea coronadoi* in scientific collections, we present herewith photographs of both examples, and some elaborations upon the description given by Woodring & Olsson (1957).

Description.—Both specimens are immature and lack the reflected outer lip of fully developed individuals. The smaller specimen, 49.5 mm in length, retains the nuclear whorls although they are somewhat weathered. There are slightly over three helicoid, inflated whorls apparently lacking sculpture, ending with a distinct axial riblet. Spiral sculpture, including the nodulose shoulder, commences on the first postnuclear whorl. Axial growth lines and weak, narrow axial swellings are distinctly visible on the second postnuclear whorl but are obliterated on the first by wear. About six flattened spiral cords, including that at the shoulder, occur early in the second postnuclear whorl, increasing in number by intercalation as the whorls enlarge. On the body whorl they are numerous, low, flattened, separated by roughly their own width. The three low spiral swellings on the body whorl of the two previously known specimens, noted by Woodring & Olsson (1957: 24) have not appeared on the present examples, although the upper one is incipient. The color is light brown with indistinct paler spiral bands and

FIGURE 13. Cassididae. *Bathygalea coronadoi* (Crosse), Sta. P-739: upper, length 62.9 mm, width 42.7 mm; lower, length 49.5 mm, width 34.8 mm.

FIGURE 14. Cassididae. *Sconsia striata* (Lamarck): upper, Sta. P-1232, height 46.6 mm, width 27.7 mm; lower, Sta. P-353, height 42.8 mm, width 24 mm.

faint axial streaks a little darker than the ground color. The larger specimen is discolored a ferruginous brown on all parts up to a major pause in growth on the body whorl.

Operculum shaped like that of *Phalium*, the free margin as in *Phalium cicatricosum* without serrations.

Genus *Sconsia* Gray

Sconsia Gray, 1847, Proc. Zool. Soc. Lond., *15*: 137.—Clench & Abbott, 1943, Johnsonia, *1*(9): 6.

This genus has been reviewed by Clench & Abbott (1943). Until now, it has been known to contain only one rather variable species in the Caribbean area, extending from the Straits of Florida southward to Barbados and Panama. Trawling operations by R/V GERDA and R/V PILLSBURY have obtained several records of this species, and the latter vessel has collected a distinctive new species of the genus.

14. *Sconsia striata* (Lamarck, 1816)
Figs. 14; 17, A; 20, D

Cassidaria striata Lamarck, 1816, Encyclopédie Méthodique, Vers, *3*: 3, pl. 405, figs. 2a, b.
Sconsia striata, Clench & Abbott, 1943, Johnsonia, *1*(9): 6, pl. 4, figs. 1-4.
—Clench, 1959, Johnsonia, *3*(39): 329, pl. 172 (photo of holotype of *Sconsia barbudensis* Higgins & Marratt).

Material Examined.—GERDA Sta. 236, Straits of Florida, W of Riding Rocks (25°15'N, 79°15'W, 384 meters, 30 January 1964; 1).—G-272, Straits of Florida, S of Gun Cay (25°28'N, 79°18'W, 384-357 m, 30 March 1964; 1).—G-390, Straits of Florida, W of Little Bahama Bank (27°19'N, 79°11'W, 247-275 m, 19 September 1964; 1).—G-503, W of Pinder Point, Grand Bahama I. (26°31'N, 78°51'W, 366 m, 4 February 1965; 1).—G-625, Straits of Florida, NW of Bimini (25°53'N, 79°19'W, 384-412 m, 29 June 1965; 1).—G-638, Straits of Florida at entrance of Northwest Providence Channel (26°05'N, 79°12'W, 238-256 m, 30 June 1965, 2 juv.).—G-725, Northwest Providence Channel (26°01'N, 79°10'W, 210-143 m, 3 August 1965; 2 broken).

Also, eleven specimens, four of them broken, from eight PILLSBURY stations in the southwestern Caribbean Sea: P-349, P-353, P-361, P-362, P-365, P-366, P-367, P-783; seven specimens, one broken, from five PILLSBURY stations off the coast of Venezuela: P-714, P-716, P-752, P-756, P-760; one from the Lesser Antilles between Grenada and Trinidad, P-849; and one and fragments from off the southwest coast of Jamaica, P-1232.

Remarks.—The shells vary considerably as to color pattern and sculpture.

15. Sconsia nephele, n. sp.
Fig. 15

Description.—Shell ovate, imperforate, rather thin and glossy. The only known specimen, 37.8 mm long, has six whorls and therefore probably is not fully grown. The spire is low, flat-sided, with only slightly impressed sutures, produced at an angle of 94°. Nuclear whorls smooth, postnuclear whorls with distinct spiral grooves, seven in number on the third, fourth, and fifth whorls. On the body whorl, the grooves near the suture and the base remain distinct, but those in the middle of the whorl become rather weak and show a somewhat irregular alternation of weaker and stronger, the total number being about 40. Besides faint axial growth lines, there are slightly raised axial riblets that become stronger toward the end of the last whorl, thus producing a distinct cancellation especially noticeable anteriorly. Last whorl with one varix; outer lip thickened, denticulate within, showing evidence of continuing growth, not reflected. Parietal wall glazed in an area clearly shown in Figure 15.

Ground color nearly uniform ochre, with faint evidence of a few localized darker axial streaks following lines of growth. Pattern consisting of eight spiral bands of darker brown (burnt ochre) regularly interrupted with squarish spots of white. Varix and outer lip marked by seven elongated, rectangular spots of darker brown (umber) which alternate with the color bands.

Operculum not seen; specimen dead, but shell in fresh condition.

Holotype.—USNM No. 700004, from PILLSBURY Sta. P-851.

Type-Locality.—PILLSBURY Sta. P-851, southwest of the island of Grenada, 11°52.8'N, 61°53.3'W, depth 18 meters, 3 July 1969.

Remarks.—Compared with *Sconsia striata* (Lamarck) of similar size, the shell is more broadly ovate, the spire lower, the spiral grooves more distant and less distinct, and the color pattern quite different from any variations recorded in the literature or observed in specimens.

16. Morum (Cancellomorum) dennisoni (Reeve)
Figs. 16; 17, B

Morum dennisoni, Clench & Abbott, 1943, Johnsonia, *1*(9): 5, pl. 4, fig. 5.
—Dance, 1966, Shell Collecting: pl. 22c; 1969, Rare Shells: 80, pl. 13a (color photo of holotype).
Morum (Cancellomorum) dennisoni, Emerson, 1967, Veliger, *9*(3): pl. 39, figs. 1a, 1b (photo of "cotype").—Dance & Emerson, 1967, Veliger, *10* (2): 91-94, pl. 12, figs. 5-7 (photo of holotype; copy of original illustration of holotype).

Material Examined.—Represented at six PILLSBURY stations, as follows: P-581, Arrowsmith Bank, Yucatan (21°05'N, 86°23'W, 146-265 m, 22

FIGURE 15. Cassididae. *Sconsia nephele*, n. sp., holotype, Sta. P-851, height 37.8 mm, width 24 mm.

May 1967; 2).—P-598, Arrowsmith Bank, Yucatan (21°07'N, 86°21'W, 155-205 m, 15 March 1968; 1).—P-734, off Isla Tortuga, Venezuela (11°01.8'N, 65°34.2'W, 68-60 m, 22 July 1968; frag.).—P-772, off Guajira Peninsula, Colombia (12°20.2'N, 71°55.1'W, 11 m, 29 July 1968; 1). —P-835, SE of Trinidad (9°36'N, 60°10'W, 48 m, 30 June 1969; 1).— P-916, Guadeloupe (16°22.2'N, 61°26.3'W, 2 m, 11 July 1969; 2).

Remarks.—In the above material, all specimens but that from P-772 are similar to the material taken by the ALBATROSS at Arrowsmith Bank, Yucatan (Sta. 2354, USNM No. 93742) and to a specimen from Barbados (USNM No. 459829). These are similar to the original specimens of *O. dennisoni* as figured by Dance (1966, 1969), Emerson (1967), and Dance & Emerson (1967), but the axial ribs are not as well developed so that the sculpture is less distinctly cancellate (Fig. 16). The shell from Sta. P-772 (Fig. 16), which is more prominently cancellate and differs in shape, appears to be specifically distinct when considered in the light of the present collection only. However, as the published figures of *M. dennisoni* show some sculptural variation, and as so few specimens are available for comparison, this specimen is not given specific recognition at the present time.

FIGURE 16. Cassididae. *Morum dennisoni* Reeve: upper, Sta. P-581, height 50.5 mm, width 29.2 mm; lower, Sta. P-772, height 34.6 mm, width 21.9 mm.

Family Oocorythidae

17. *Oocorys sulcata* Fischer, 1883

Figs. 18; 19; 20, B, C; 22, B

Oocorys sulcata Fischer, 1883, Journ. Conchyl., *31*: 392.—Watson, 1886, Rep. Sci. Res. Challenger, Zool., *15*: 412, pl. 17, fig. 11.—Locard, 1897, Expéd. Scient. Travailleur et Talisman, Mollusques Testacés, *1*: 288.—Turner, 1948, Johnsonia, *2*(26): 186, pl. 75, fig. 8 (operculum); pl. 85, figs. 1-2 (shells).

Oocorys sulcata var. *minor* Locard, 1897, Expéd. Scient. Travailleur et Talisman, Mollusques Testacés, *1*: 290.

Oocorys sulcata var. *elongata* Locard, 1897, Expéd. Scient. Travailleur et Talisman, Mollusques Testacés, *1*: 291.

Records.—EASTERN ATLANTIC: PILLSBURY stations P-18 (5°01'N, 00°12'E, 3047-3129 m, 26 May 1964; 3).—P-44 (5°05'N, 4°00'W, 586-403 m, 30 May 1964; 3).

WESTERN ATLANTIC: PILLSBURY stations P-325 (9°52'N, 79°35.5'W, 1774-1656 m, 7 July 1966; 1).—P-346 (9°54.5'N, 77°03'W, 2950-2938 m; 1).—P-374 (9°57'N, 76°10.6'W, 434-373 m, 14 July 1966; 1).—P-586 (23°32'N, 82°33'W, 1737-1682 m, 24 May 1967; 1).—P-680 (8°42'N, 55°48'W, 3014-3239 m, 13 July 1968; 2).—P-681 (8°11.5'N, 56°12'W, 2672-2736 m, 14 July 1968; 3).—P-719 (11°35'N, 64°35.5'W, 1409-1629 m, 20 July 1968; 1).—P-748 (11°24.8'N, 67°10.1'W, 1867-1784 m, 25 July 1968; 14).—P-844 (11°30'N, 60°14.5'W, 1464-1848 m, 1 July 1969; 2)—P-1178 (19°14'N, 73°14'W, 1798-1902 m, 30 June 1970; 6).—P-1180 (18°55'N, 73°53'W, 3493-3109 m, 1 July 1970; 4).

Remarks.—Specimens from the West Atlantic are indistinguishable from West African material on conchological grounds, but the radula has one or two conspicuous sharp denticles on the inside of the first (inner) marginal instead of one to three inconspicuous bumps, and both the lateral and rachidian teeth have fewer denticles. Whether these differences are constant, indicating a small degree of differentiation, is impossible to determine now, as so little material has so far been studied. The taxonomic significance of the horny jaws has not been investigated.

Genus *Dalium* Dall, 1889

Dalium Dall, 1889a, Bull. Mus. comp. Zool. Harv., *18*: 230.—Thiele, 1929, Handb. syst. Weichtierkunde, *1*: 279.—Clench & Abbott, 1943, Johnsonia, *1*(9): 8.

As the operculum and radula were unknown at the time of its original establishment by Dall, this genus was only tentatively placed in the family Oocorythidae. Thiele treated it in the family Cassididae, as did Clench & Abbott, who did not recognize the Oocorythidae as a distinct family but

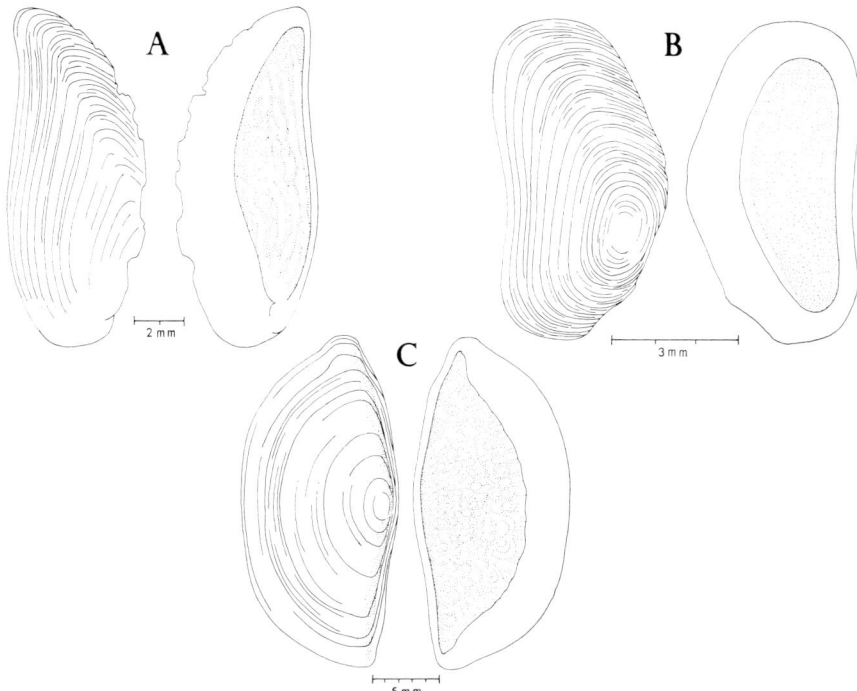

FIGURE 17. Cassididae. Opercula: A, *Sconsia striata* (Lam.), Sta. P-783; B, *Morum dennisoni* Reeve, Sta. P-916; C, *Bathygalea coronadoi* (Crosse), Sta. P-739. Drawings by Constance S. McSweeny.

ranked it as a subfamily of the Tonnidae. Olsson (1964:171) recognized the family Oöcoritidae (sic) and assigned *Dalium* to it. He reported the common occurrence of a late Neogene fossil *Dalium* much like *D. solidum* from the Esmeraldas Formation of Ecuador.

Material obtained by R/V PILLSBURY at several localities in the southern Caribbean Sea shows that the radula (Fig. 20, A) is virtually indistinguishable from that of *Oocorys* (Fig. 20, B, C), and that the operculum is weakly spiral with a terminal nucleus (Fig. 22, A), closer to that of *Oocorys* (Fig. 22, B) than to those of cassidids with central or lateral nucleus. Hence, the relationship originally deduced by Dall can be supported, even though the ultimate status of the Oocorythidae may remain open to discussion.

Type-Species.—*Dalium solidum* Dall, by original designation.

FIGURE 18. Oocorythidae. *Oocorys sulcata* Fischer: upper, Sta. P-681, western Atlantic, height 33.5 mm, width 26.8 mm; lower, Sta. P-18, eastern Atlantic, height 44.7 mm, width 33 mm.

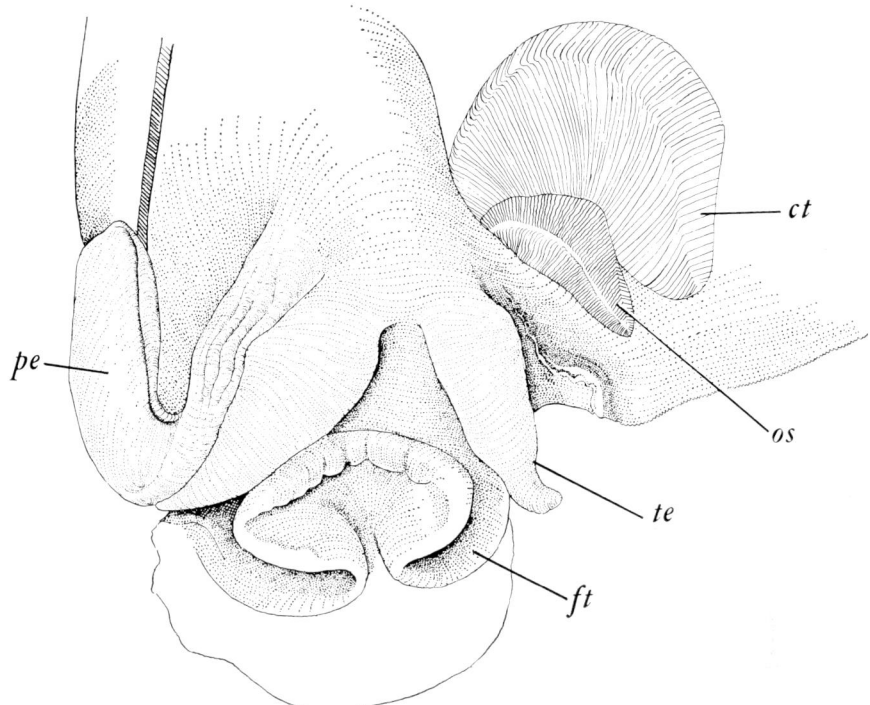

FIGURE 19. Oocorythidae. *Oocorys sulcata* Fischer, gross anatomy of mantle cavity of male from Sta. P-18. (*ct*, ctenidium; *ft*, foot; *os*, osphradium; *pe*, penis; *te*, tentacle.)

18. *Dalium solidum* Dall, 1889
Figs. 20, A; 21; 22, A

Dalium solidum Dall, 1889a, Bull. Mus. comp. Zool. Harv., *18*: 230, pl. 19, fig. 10d (off Grenada, Lesser Antilles, 576 fathoms, BLAKE Sta. 265).— Clench & Abbott, 1943, Johnsonia, *1*(9): 8, pl. 4, figs. 6-7.

Material Examined.—Fifty-four specimens from the southern Caribbean off Panama, Colombia, Venezuela, and Surinam, at PILLSBURY Stas. P-381 (10°17.0'N, 75°59.9'W, 733-604 m; 1).—P-447 (9°07.4'N, 81°07.4'W, 664-681 m; 6).—P-448 (9°10.1'N, 80°55.6'W, 962-878 m; 2).—P-672 (07°37'N, 55°22'W, 1336-1221 m; 1).—P-682 (07°33.5'N, 56°25.0'W, 1318-1345 m; 2).—P-719 (11°35.0'N, 64°34.5'W, 1409-1629 m; 30).— P-747 (11°46.0'N, 67°05.7'W, 1175-1098 m; 1).—P-748 (11°24.8'N, 67°10.1'W, 1867-1784 m; 4).—P-754 (11°36.9'N, 68°42.0'W, 684-1574 m; 1).—P-770 (12°55.0'N, 71°46.5'W, 1318-1299 m; 6).

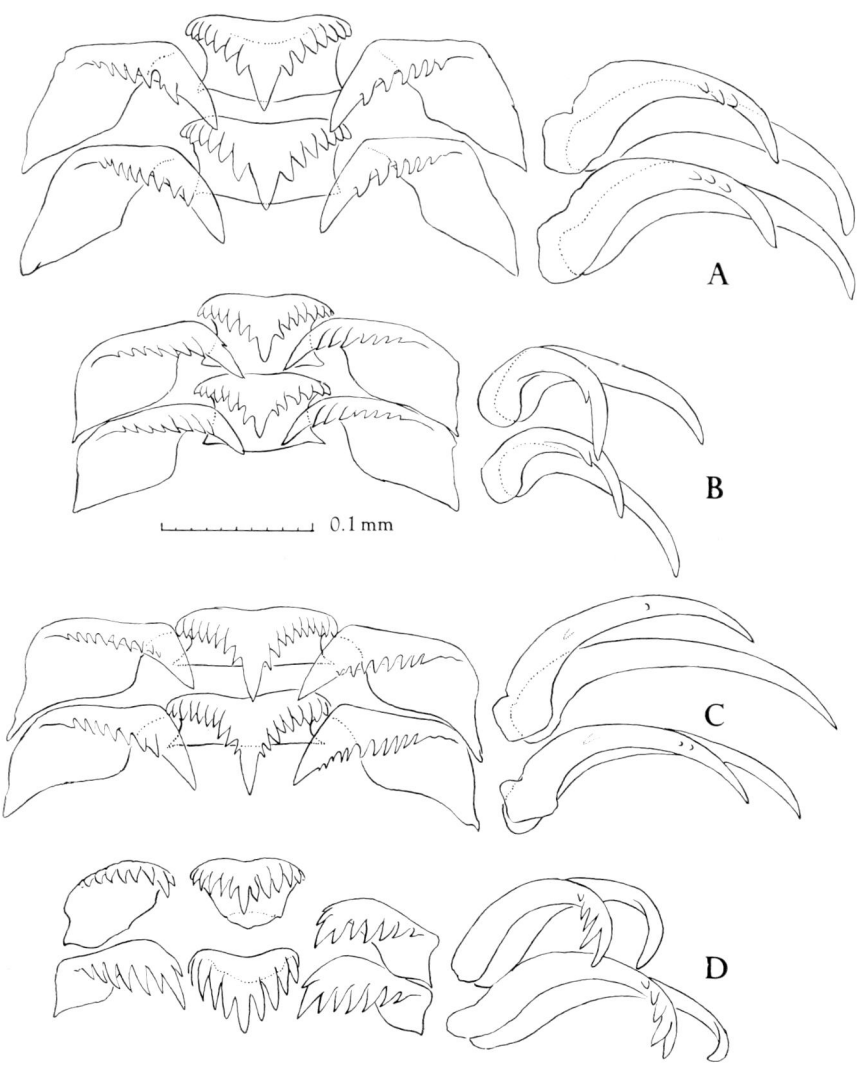

FIGURE 20. Radulae: A, *Dalium solidum* Dall, Sta. P-682.—B-C, *Oocorys sulcata* Fischer: B, Sta. P-681; C, Sta. P-18.—D, *Sconsia striata* (Lamarck), Sta. P-783. All figures drawn to same scale.

FIGURE 21. Oocorythidae. *Dalium solidum* Dall, Sta. P-719, height 58.6 mm, width 33 mm.

Remarks.—It is noteworthy that all stations that yielded *D. solidum* reached depths exceeding 1000 meters, except those in the Gulf of Uraba (P-381, P-447, and P-448); all the specimens from the deepest station (P-748, 1784-1867 meters) were dead and badly damaged by erosion and breakage.

Dall's type-specimen measured 41.25 mm in length and 23.00 mm in width. The present material ranges from a length of 23.1 mm and breadth of 15.0 mm (P-447) to a length of 64.0 mm and a breadth of 37.1 mm (P-719).

Dall stated that the periostracum, "if any, is extremely thin," and Clench & Abbott (1943) after reexamining the same material concluded that there was "no periostracum apparent." In all of the present live-collected material, there is a thin, olivaceous periostracum worn smooth on top of the spiral cords but showing distinct, erect axial lamellae between them and everywhere in the region just behind the outer lip where abrasion has not

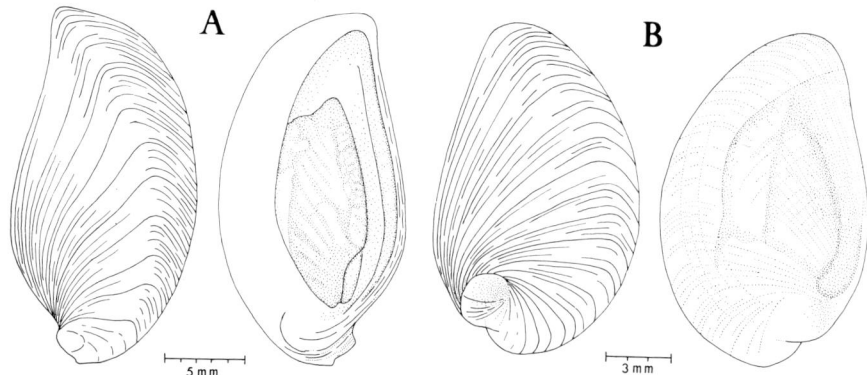

FIGURE 22. Opercula: A, *Dalium solidum*, Sta. P-682; B, *Oocorys sulcata*, P-681. Drawings by Constance S. McSweeny.

yet affected them. There is a tendency for the periostracum to be worn away in the subsutural channel, leaving a white spiral band ascending the spire. Beneath the periostracum, the shell is chalky white. The early whorls are generally much eroded, and even in the smallest specimens only the shelly filling of the nuclear whorls remains.

The operculum (Fig. 22, A) is ovate-trigonal, weakly spiral, with terminal nucleus, not so strongly spiral as that of *Oocorys* (Fig. 22, B) but not lateral as in *Phalium, Bathygalea* (Fig. 17, C), *Sconsia* (Fig. 17, A) and *Cassis*.

The radula of *Dalium* (Fig. 20, A) is essentially as in *Oocorys*; the marginal teeth are not so long and slender as in *Cassis* (see Thiele, 1929: 278, fig. 295a), and the inner marginal does not have the strong apical denticulation as in *Sconsia* (Fig. 20, D).

Family Ficidae
Genus *Ficus* Röding, 1798

Specimens of *Ficus* with shells spotted as in *F. howelli* Clench & Aguayo and *F. atlantica* Clench & Aguayo were taken by R/V PILLSBURY at five stations along the northern coast of South America, both inside and outside the Caribbean Sea, and on the southeastern coast of Hispaniola. Some of these have widely spaced reddish brown dots in six spiral rows as in *F. howelli*, whereas others have the spots more closely spaced in 9-12 spiral rows as in *F. atlantica*. The largest specimens are those with close spotting, the reverse of the situation with the original specimens described by Clench

& Aguayo (1940). Minor sculptural differences are present but do not appear to be correlated with color pattern.

Eleven living specimens were obtained, five males and six females. All males have six spiral rows of major spots (Fig. 23, A), two of them with some weaker intermediate spots in imperfect spirals. Five of the six females have nine or more spiral rows of dots (Fig. 23, C, D), but one has six only. The latter is a small example, 31.6 mm long. In one of the other female specimens it can be seen that the rows of dots increased rather abruptly from six to nine (Fig. 23, B). It appears that the spotting is similar in young shells of both sexes, becoming denser in females with the approach of maturity.

The curvature of the columella and the ventricosity of the shell also vary somewhat, but neither seems correlated with sex or color pattern. In the seven larger examples, presumably adults, the percentage of the diameter in the total length is 55, 56, 58, and 59 for the females, and 54, 59, and 59 for the males. Thus, both sexes have about the same range in ventricosity.

It seems clear that *Ficus howelli* Clench & Aguayo and *F. atlantica* Clench & Aguayo represent variations of a single species. The color pattern of their shells suggests that *F. howelli* is the male and *F. atlantica* the female. These nominal species are now united under the name:

19. *Ficus howelli* Clench & Aguayo
Figs. 23, 59

Ficus howelli Clench & Aguayo, 1940, Mem. Soc. Cub. Hist. nat., *14*(1): 85, pl. 14, fig. 2.—Clench, 1945, Johnsonia, *1*(18): 1, pl. 1, figs. 1-2 (off Bahía de Cochinos, Sta. Clara, Cuba, 175–225 fms = 315-405 m).

Ficus atlanticus Clench & Aguayo, 1940, Mem. Soc. Cub. Hist. nat., *14*(1): 85, pl. 14, fig. 1.—Clench, 1945, Johnsonia, *1*(18): 2, pl. 1, figs. 3-4 (off São Salvador, Brasil, 450 fms = 810 m).

Material Examined.—Ten specimens from five PILLSBURY stations, as follows: P-650, off French Guiana (06°07.0′N, 52°19.0′W, 84-91 m, 8 July 1968; 1 ♀).—P-728, Gulf of Cariaco, Venezuela (10°22.5′N, 65°23.0′W, 86 m, 21 July 1968; 2 ♀ ♀).—P-737, off Cabo Codera, Venezuela (10°44.0′N, 66°07.0′W, 60-73 m, 22 July 1968; 2 ♂ ♂).—P-838, off east coast of Trinidad (10°32.0′N, 60°23.0′W, 93-115 m, 30 June 1969; 2 ♂ ♂, 1 ♀).—P-1303, off SE coast of Hispaniola (18°21′N, 69°14.3′W, 172 m, 21 July 1970; 1 ♂, 2 ♀ ♀).

The largest specimen is a female with a shell 46.5 mm long, 26.7 mm wide (Sta. P-1303); the smallest specimen is a female with a shell 20 mm long, 11 mm wide (Sta. P-728); the largest male has a shell 39.3 mm long, 22.5 mm wide (Sta. P-838).

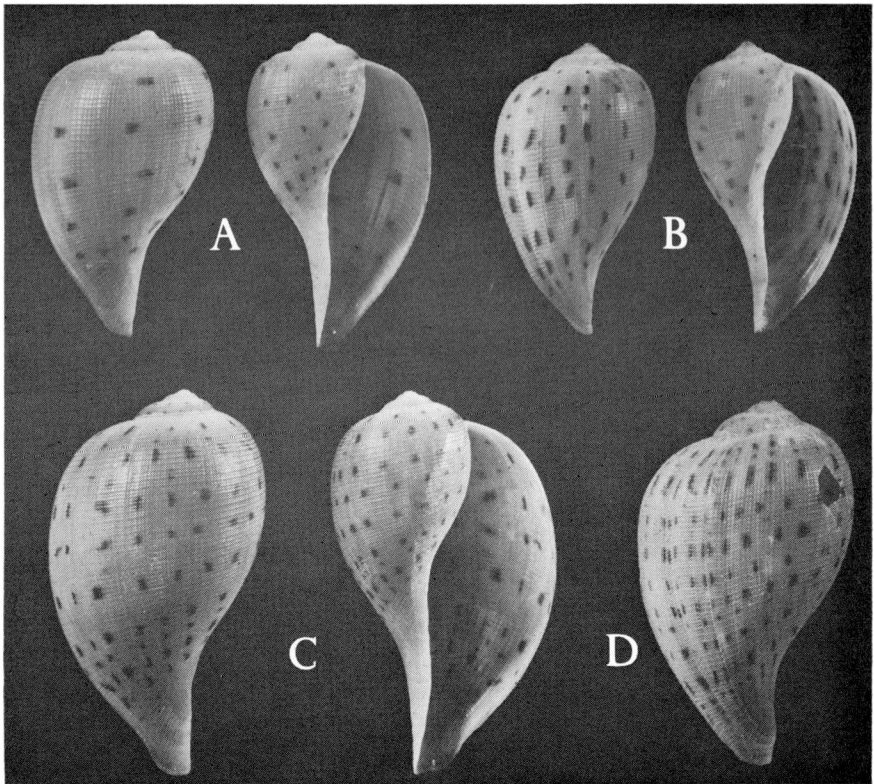

FIGURE 23. Ficidae. *Ficus howelli* Clench & Aguayo: A, male, Sta. P-1303, length 37.6 mm, width 22.2 mm; B, female, Sta. P-728, length 35.4 mm, width 20.5 mm; C, female, Sta. P-1303, length 46.5 mm, width 26.7 mm; D, female, Sta. P-650, length 44.8 mm, width 26.2 mm.

Diagnosis.—Shell pyriform, thin, rather small (maximal length probably not exceeding 50 mm), light yellowish brown in color with 6-12 spiral rows of small reddish brown dots, which are more widely spaced in males (Fig. 23, A) than in females (Fig. 23, B-D). Whorls about five when shell is fully grown, 4½ to 4¾ in specimens approximately 35 mm in length. Sculpture closely cancellate, the spiral cords increasing in number by intercalation.

Remarks.—*Ficus pellucida* Deshayes (1856, Journ. Conchyl., 5: 184, pl. 6, figs. 1-2; "Sa patrie nous est inconnue") agrees in all respects except size —70 mm.

FIGURE 24. Muricidae. *Murex olssoni* Vokes, Sta. P-366, length 51 mm.

Family Muricidae
20. *Murex (Murex) olssoni* Vokes
Fig. 24

Murex (Murex) olssoni Vokes, 1967, Tulane Stud. Geol., 5(2): 84, pl. 3, figs. 1-3.

Description.—See Vokes, 1967.

Holotype.—USNM No. 677704.

Type-Locality.—Gulf of Morrosquillo, Colombia, OREGON Sta. 4896: 9°36'N, 75°52.5'W, 23-27 fathoms = 42-49 meters, 26 May 1964.

Records.—Taken at the following thirty stations of R/V PILLSBURY between Golfo de los Mosquitos, Panama, and Cartagena, Colombia, all in July, 1966: P-324 (9°44.0'N, 79°31.0'W, 64-55 m; 6).—P-330 (9°37.5'N, 78°54.0'W, 128-64 m; 1).—P-332 (9°31.2'N, 78°53.0'W, 51 m; 3).—

P-333 (9°33.0'N, 78°49.0'W, 57 m; 11).—P-334 (9°33.0'N, 78°50.0'W, 51 m, 3).—P-347 (8°43.0'N, 77°03.0'W, 55-53 m; 2)—P-362 (8°57.5'N, 76°33.6'W, 64-55 m; 8).—P-365 (9°31.1'N,76°15.4'W, 56-58 m; 28).— P-366 (9°31.0'N, 75°59.5'W, 37-33 m; 7).—P-367 (9°31.1'N, 75°49.6'W, 37-35 m; 18).—P-368 (9°31.2'N, 75°41.1'W, 37 m; 7).—P-369 (9°35.7' N, 75°37.6'W, 18 m; 1).—P-370 (9°37.9'N, 75°50.4'W, 37 m; 8).—P-371 (9°40.0'N, 76°01.5'W, 46-55 m; 4).—P-379 (10°02.2'N, 75°41.3'W, 55 m; 1).—P-389 (9°53.8'N, 75°50.9'W, 70-51 m; 2).—P-392 (9°45.1'N, 76°09.1'W, 79-75 m; 2).—P-396 (9°18.2'N, 76°24.8'W, 70-68 m; 6).— P-397 (9°12.8'N, 76°27.1'W, 62-66 m; 5).—P-402 (8°51.2'N, 77°01.6'W, 73 m; 6).—P-403 (8°48.7'N, 77°12.7'W, 99-97 m; 1).—P-405 (8°49.2'N, 77°21.2'W, 92-93 m; 1).—P-411 (8°40.7'N, 77°21.8'W, 29-42 m; 4).— P-412 (8°38.9'N, 77°13.2'W, 55-60 m; 9).—P-425 (9°38.9'N, 79°15.3' W, 70-64 m; 4).—P-432 (9°18.2'N, 80°03.3'W, 24 m; 9).—P-433 (9° 20.5'N, 80°13.5'W, 68-64 m; 1).—P-435 (9°08.5'N, 80°29.5'W, 37-48 m; 9).—P-437 (9°00.1'N, 80°45.8'W, 55 m; 16).—P-450 (9°22.0'N, 79° 56.0'W, 9 m; 3).

Remarks.—It is interesting to note that this species, described as late as 1967, was found to be the commonest and most widespread shallow-water muricid along the coast of Panama and Colombia. A gap in its distribution, between Cabo de San Blas and Cabo Tiburon, may be more apparent than real, as the Continental Shelf there is so narrow and rough that sampling may not have been representative.

21. *Murex (Murex) donmoorei* Bullis
Fig. 25

Murex donmoorei Bullis, 1964, Tulane Stud. Zool., *11*(4): 101, figs. 1-2.

Holotype.—USNM No. 635146.

Type-Locality.—45 miles north of St. Andrews Point, British Guiana; OREGON Sta. 2254: 7°07'N, 57°08'W, 20-22 fm (= 37-40 m).

Records.—Taken at 16 PILLSBURY stations between French Guiana and Paraguana Peninsula, Venezuela: P-648 (5°26'N, 52°12'W, 42 m, 8 July 1968; 6).—P-650 (6°07'N, 52°19'W, 84-91 m, 8 July 1968; 2).—P-653 (6°12'N, 52°58'W, 9 July 1968; 3).—P-669 (6°39'N, 55°15'W, 33 m, 10 July 1968; 4).—P-686 (7°00'N, 57°08'W, 27-26 m, 15 July 1968; 3).— P-695 (8°12'N, 58°33'W, 37 m; 15 July 1968; 1).—P-714 (11°29'N, 63°24.3'W, 59 m, 19 July 1968; 1).—P-718 (11°22.5'N, 64°08.6'W, 60 m, 20 July 1968; 4).—P-721 (11°06.5'N, 64°22.5'W, 26-27 m, 21 July 1968; 4).—P-723 (10°43.5'N, 64°16'W, 71-60 m, 21 July 1968; 1).— P-731 (10°20'N, 65°41'W, 57-60 m, 22 July 1968; 4).—P-749 (10°37'N, 67°57.9'W, 59 m, 25 July 1968; 1).—P-750 (10°36.1'N, 68°12.2'W, 22-

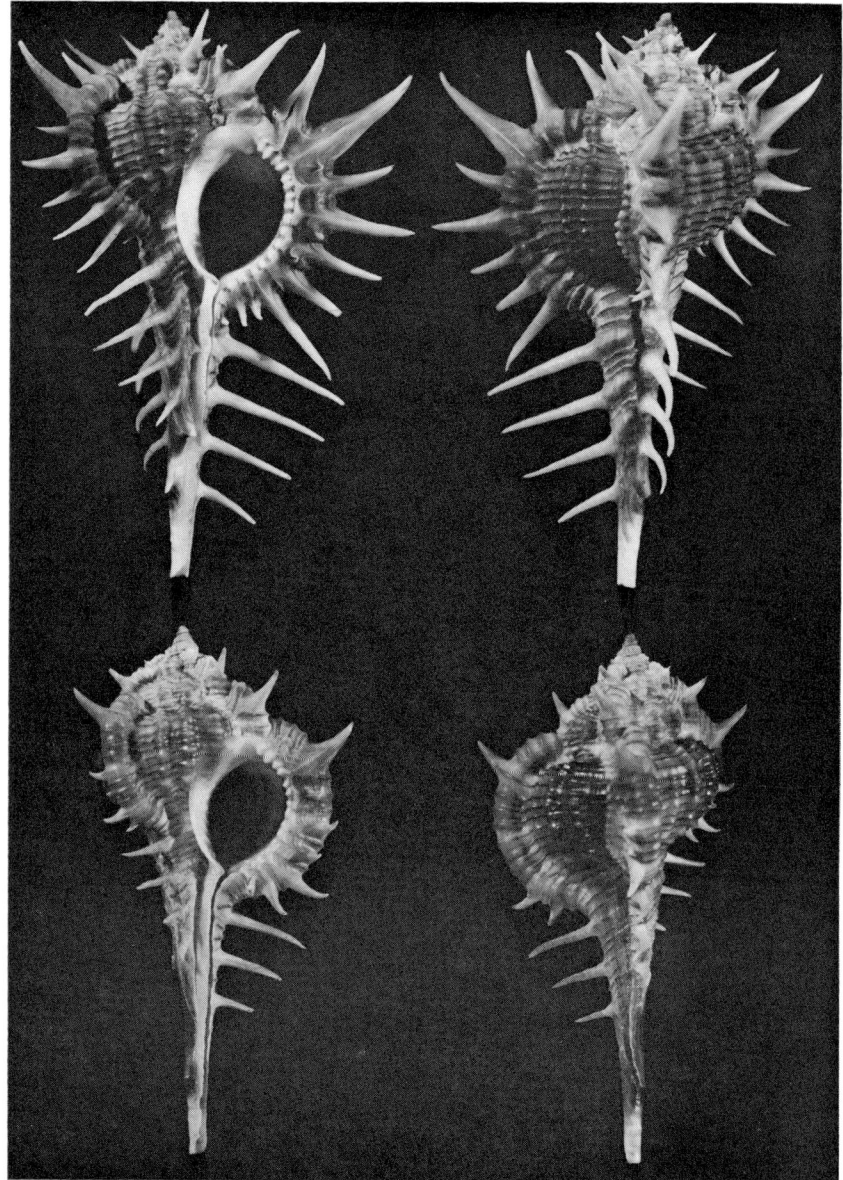

FIGURE 25. Muricidae. *Murex donmoorei* Bullis: upper, Sta. P-749, length 65.4 mm; lower, Sta. P-695, length 51.6 mm.

26 m; 25 July 1968; 15).—P-756 (11°33.1′N, 69°12.6′W, 16-38 m, 27 July 1968; 1).—P-758 (11°42.2′N, 69°40′W, 15-18 m, 27 July 1968; 1). —P-761 (11°52′N, 70°22′W, 35 m, 27 July 1968; 1).

Remarks.—As the present material conforms in essential details with the original description and with the type-specimens (holotype, USNM No. 635146; paratype, UMML 30-2770) that were available for comparison, the 16 lots enumerated above are referred to *M. donmoorei.* All have a large, carinate nucleus, a straight siphonal canal of only moderate length bearing three or four (rarely two) spines, the varices usually have six spines, and the spiral cords are marked by a brown line.

The specimens in the PILLSBURY collection vary in the intensity of coloration and the distinctness of the spiral lines, length of spines, length of anterior canal and, most significantly, in the presence or absence of a strong spine between the shoulder spine and the suture. Moreover, the holotype has this spine well developed (Bullis, 1964: figs. 1, 2), but the paratype (UMML 30-2770) lacks it, so it is possible that two species are involved in the type-series. However, pending a thorough investigation, all the specimens are retained under the name *donmoorei.*

The unusually elaborate specimen from Sta. P-749 shown in Figure 25 agrees with the holotype in the presence of the strong spine between shoulder and suture and is thus in best agreement with *Murex donmoorei.* All other specimens lack this spine as shown (Fig. 25), in agreement with the paratype, and therefore may eventually be referred to another species. They resemble *M. olssoni* Vokes in general appearance but in that species the nucleus is smaller and not carinate, the spire is higher, the spiral cords generally are not marked by brown lines, and the anterior canal usually has only one or two spines, rarely three.

22. *Murex* (*Murex*) *cabritii* Bernardi
Fig. 26

Murex cabritii Bernardi, 1858, Journ. Conchyl., 7: 301, pl. 10, fig. 3
Murex (*Murex*) *cabritti,* Clench & Pérez Farfante, 1945, Johnsonia, *1*(17): 3, pl. 1 (synonymy).

Records.—This species has been taken at the following stations by R/V PILLSBURY: P-574, NE of C. Gracias a Dios, Nicaragua (16°16′N, 82°26.5′W, 37 m, 20 May 1967; 3).—P-615, Gulf of Honduras (16°01.5′N, 88°42.5′W, 13 m, 19 March 1968; 4).—P-616, Gulf of Honduras (16°01′N, 88°43′W, 13 m, 19 March 1968; 1).—P-623, off Cabo de Honduras: 16°00′N, 86°08′W, 42-55 m, 21 March 1968; 3).—P-625, off Cabo de Honduras (15°59.5′N, 86°02.5′W, 27-37 m, 21 March 1968; 5).—P-626, off Cabo de Honduras (15°57.6′N, 86°09′W, 35-40 m, 21 March 1968; 3).

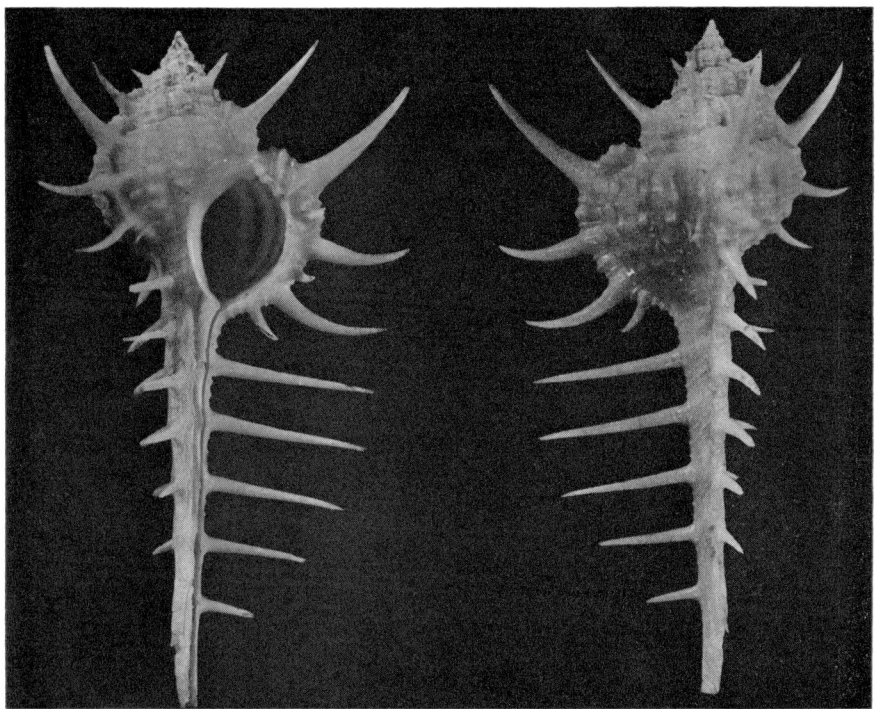

FIGURE 26. Muricidae. *Murex cabritii* Bernardi, Sta. P-615, length 61.7 mm.

—P-627, off Cabo de Honduras (15°56.5'N, 86°14.0'W, 46 m, 21 March 1968; 3).

Remarks.—A photograph of a specimen from Sta. P-615 is given for comparison with *Murex donmoorei* Bullis (Fig. 25).

23. *Murex* (*Siratus*) *beauii* Fischer & Bernardi
Fig. 27

Murex beauii Fischer & Bernardi, 1857, Journ. Conchyl., 5: 295, pl. 3, fig. 1.
Murex (*Murex*) *beauii*, Clench & Pérez Farfante, 1945, Johnsonia, *1*(17): 14, pl. 7.
Murex (*Murex*) *branchi* Clench, 1953, Johnsonia, *2*(32): 360, pl. 179.
Murex (*Siratus*) *beauii*, Bullis, 1964, Tulane Stud. Zool., *11*(4): 104.

Records.—This species has been taken at several stations in the Straits of Florida and Antilles by R/V GERDA and R/V PILLSBURY. Although the records will not be enumerated in detail at this time, the following are representative: GERDA Sta. G-1036, Straits of Florida SE of Marquesas

FIGURE 27. Muricidae. *Murex beauii* Fischer & Bernardi, Sta. P-876; upper, webbed form, length 49.5 mm; lower, "*branchi*" form, length 95.5 mm.

Keys (24°22.5'N, 80°53'W, 229-238 m, 26 February 1969; one specimen, *branchi* form).—GERDA Sta. G-1082, Straits of Florida SE of Sombrero Key (24°24.5'N, 82°02.5'W, 115 m, 26 April 1969; one specimen, *branchi* form).—PILLSBURY Sta. P-876, Lesser Antilles, off St. Vincent (13°13.9'N, 61°04.7'W, 231-258 m, 6 July 1969; five webbed specimens, one *branchi* form).—PILLSBURY Sta. P-943, Lesser Antilles, N of Guadeloupe (16°25.9'N, 61°36.7'W, 275 m, 17 July 1969; six webbed specimens).

Remarks.—As already suggested by Vokes (1963: 111), *Murex branchi* Clench is nothing but *M. beauii* without elaborate varical frills. The *branchi* form appears to be the usual form off the Florida Keys, but in the Antilles it occurs along with webbed examples. The specimens illustrated in Figure 27 are from the same haul (P-876).

24. *Murex (Chicoreus) brevifrons* Lamarck
Figs. 28, 29

Murex brevifrons Lamarck, 1822, Hist. Nat. Animaux sans Vertèbres, 7: 161.
Chicoreus (Chicoreus) brevifrons, E. Vokes, 1965, Tulane Stud. Geol., 3(4): 192, pl. 3, fig. 5 (synonymy).

Records.—Taken by R/V PILLSBURY at the following stations: P-648, French Guiana (05°26.0'N, 52°12.0'W, 41 m, 8 July 1968; 3 juv.).—P-650, French Guiana (06°07.0'N, 52°19.0'W, 83-91 m, 8 July 1968; 3 juv.).—P-655, French Guiana (06°07.0'N, 53°39.0'W, 25 m, 9 July 1968; 2 adults, 5 juv.).—P-663, Surinam (06°29.0'N, 54°41.0'W, 23 m, 10 July 1968, 1 juv.).—P-711, Venezuela (10°48.0'N, 63°13.0'W, 38-40 m, 19 July 1968; 3 adults).—P-721, Venezuela (11°06.5'N, 64°22.5'W, 25-27 m, 21 July 1968; 1 adult).—P-750, Venezuela (10°36.1'N, 68°12.2'W, 22-25 m, 25 July 1968; 2 juv.).—P-758, Venezuela (11°42.2'N, 69°40.0'W, 14-18 m, 27 July 1968; 2 adults).—P-761, Venezuela (11°52.0'N, 70°22.0'W, 34 m, 27 July 1968; 1 juv.).—P-772, Colombia (12°20.2'N, 71°55.1'W, 11 m, 29 July 1968; 1 juv.).

Remarks.—Some of the variations of *M. brevifrons* are so extreme that they hardly can be recognized as belonging to the species without comparison of intermediate forms. R/V PILLSBURY obtained such specimens at Sta. P-650, which can be linked by intergrades to the more usual growth form. The largest of these, which superficially resembles *M. argo* Clench & Pérez Farfante, is illustrated (Fig. 29), together with intermediate and typical forms (Fig. 28).

25. *Murex (Paziella) actinophorus* (Dall)
Figs. 30; 35, D

Trophon (Boreotrophon) actinophorus Dall, 1889a, Bull. Mus. comp. Zool. Harv., *18*: 206, pl. 15, fig. 2 (Dall's recognition of *Boreotrophon* at sub-

FIGURE 28. Muricidae. *Murex brevifrons* Lamarck; upper, Sta. P-721, length 102.2 mm; lower, Sta. P-655, length 79.4 mm.

FIGURE 29. Muricidae. *Murex brevifrons* Lamarck, Sta. P-650, length 62.1 mm.

generic level is clearly indicated on p. 18 in the systematic list at the beginning of his paper); 1889b, Bull. U. S. natn. Mus., *37*: 120, pl. 15, fig. 2 name only; figure copied from 1889a). —M. Smith, 1939, Illustr. Catalogue Rock Shells: 19, pl. 9, fig. 3 (name only; figure copied from Dall). *Trophon actinophorus*, Bullis, 1964, Tulane Stud. Zool., 11(4): 107 (range extended to northern Brazil).

Description.—Dall, 1889a: 206.

Material Examined.—Twenty-eight specimens from the following stations of R/V GERDA and R/V PILLSBURY: G-524, Northwest Providence Channel (26°17'N, 78°41'W, 512-713 m, 3 March 1965; 1).—P-340, eastern Panama (9°13.5'N, 77°46'W, 307-366 m, 9 July 1966; 5).—P-394, off Golfo de Morrosquillo (9°28.6'N, 76°26.3'W, 421-641 m, 16 July 1966; 1).—P-445, Golfo de los Mosquitos (9°02.3'N, 81°23.8'W, 342-346 m, 21 July 1966; 2).—P-861, off the Grenadines (12°42'N, 61°05.5'W, 18-744 m, 4 July 1969; 1).—P-889, off St. Lucia (14°04.4'N, 60°50.8'W, 371-403 m, 7 July 1969; 1).—P-906, off Martinique (14°26.5'N, 60°59.2'W, 274-338 m, 9 July 1969; 1).—P-984, NW of Anguilla (18°26.4'N,

FIGURE 30. Muricidae. *Murex actinophorus* Dall, Sta. P-984: upper, two views of specimen 27.8 mm long; lower, three views of specimen 23.0 mm long.

63°12.6'W, 393-451 m, 22 July 1969; 2).—P-1225, south of Jamaica (17°42.5'N, 77°58'W, 549-530 m, 6 July 1970; 14).

Remarks.—Although the radula (Fig. 35, D) is similar to that of species of *Trophon*, it also resembles that of *Typhis* and, as in that genus, has an accessory semicircular chitinous jawlike structure that survives treatment

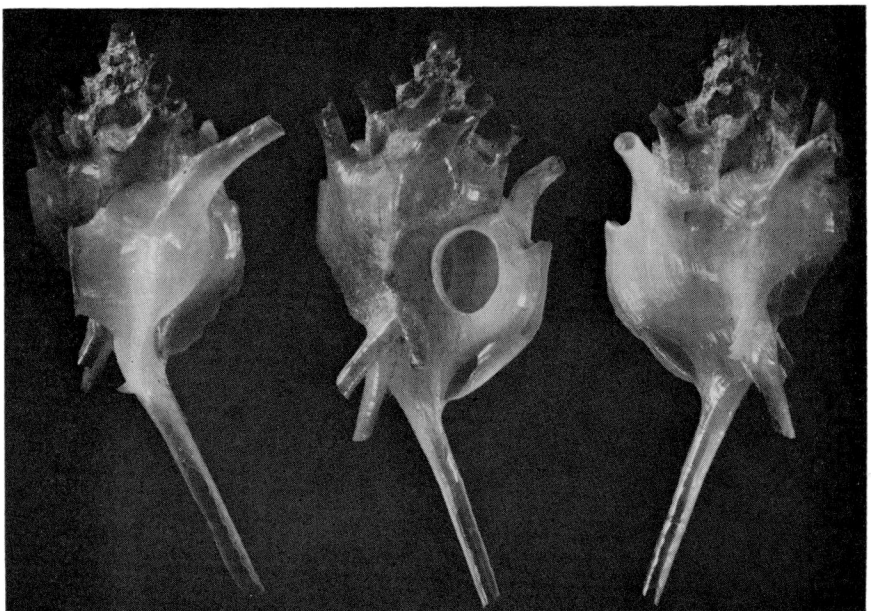

FIGURE 31. Muricidae. *Typhis longicornis* Dall, Sta. G-720, length 17.7 mm.

with KOH. Most significantly, the radula of *"Trophon" actinophorus* is almost indistinguishable from that of *Murex* (*Paziella*) *pazi* Crosse, which also has the jawlike structure as in *Typhis* but without a central cluster of denticles. As the shell of *"Trophon" actinophorus* is muricoid rather than trophonoid and, indeed, differs from *M.* (*Paziella*) *pazi* chiefly by the absence of spines around the base of the body whorl and by the thin outer lip which apparently does not develop denticles even in large examples, the species is here considered in the subgenus *Paziella*. The relationship of *Paziella* and *Poirieria* cannot be discussed at this time, but the two certainly are very close (see Vokes, 1964).

Originally described from BLAKE stations off St. Croix, Martinique, and Barbados, the range of this species was extended southward to the mouth of the Amazon River by Bullis (1964). The illustrated specimens from Sta. P-984, the larger of which is 27.8 mm in length, confirm Dall's suspicion that his specimen 17.5 mm in length was immature.

26. *Typhis* (*Siphonochelus*) *longicornis* Dall
Figs. 31; 34, A; 35, A; 36, B

Typhis longicornis Dall, *in* Agassiz, 1888, Bull. Mus. comp. Zool. Harv., *15* (2): 70, fig. 294; 1889b, Bull. U. S. natn. Mus., *37*: 122, pl. 15, fig. 7; pl. 38, fig. 15 (listed only).

Typhis (Trubatsa) longicornis, Dall, 1889a, Bull. Mus. comp. Zool. Harv., *18*: 216, pl. 15, fig. 7; pl. 38, fig. 5.—Smith, 1939, Illustr. Cat. Rock Shells: 19, pl. 14, fig. 10 (photo of Dall's larger specimen).
Siphonochelus (Siphonochelus) longicornis, Keen, 1944, J. Paleont. *18*(1): 58, 65.—Gertman, 1969, Tulane Stud. Geol., *7*(4): 170.

Material Examined.—From GERDA stations G-524, Northwest Providence Channel (26°17'N, 78°41'W, 513-715 m, 3 March 1965; 2).—G-678, Northwest Providence Channel (25°57'N, 78°13'W, 540-576 m, 20 July 1965; 1).—G-720, W entrance of Northwest Providence Channel (26°22'N, 79°11'W, 476-500 m, 3 August 1965; 3).—G-721, W entrance to Northwest Providence Channel (26°23'N, 79°04'W, 494-487 m, 3 August 1965; 1).—G-722, W entrance to Northwest Providence Channel (26°15'N, 78°57'W, 393-392 m, 3 August 1965; 1).—G-935, N of Little Bahama Bank, 27°37'N, 78°52'W, 466-417 m, 30 September 1966; 1).—G-1008, Santaren Channel (24°03'N, 79°36'W, 540-576 m, 4 June 1968; 1).—G-1012, Santaren Channel (23°35'N, 79°33'W, 508-530 m, 14 June 1968; 1).—G-1015, Santaren Channel (23°34'N, 79°17'W, 525-516 m, 15 June 1968; 1).—G-1018 (24°07'N, 79°28'W, 546 m, 15 June 1968; 1).

PILLSBURY stations P-1225, south of Jamaica (17°42.5'N, 77°58'W, 549-530 m, 6 July 1970; 2).—P-1255, southwest of Jamaica (17°18'N, 78°32'W, 805-722 m, 14 July 1970; 1).

Remarks.—The specimens taken at the 12 stations listed above clearly represent Dall's species, although they are mostly not quite so slender as the types. A good size-range is represented, but all the present specimens are white; the larger ones do not have areas of pale rosy brown as reported by Dall. Moreover, according to Dall's description, the varices are "not fimbriated, with rounded edges," but the holotype (USNM No. 94780) and two paratypes (No. 94781) clearly show the broken edge of a thin varical expansion. In all specimens taken by R/V PILLSBURY and R/V GERDA, there also is clear indication of a thin, capelike expansion from the rounded varices. This expansion, especially on the penultimate and earlier varices, is more or less broken in all specimens. In those that evidently have been in resting condition for some time, it is broken away completely even on the final varix, but it leaves a scar which, once recognized, is clearly distinguishable. In the most nearly perfect specimens, this varical expansion is conspicuous and forms a projection in front of the tube which corresponds to the varical spine of *Siphonochelus (Laevityphis) bullisi* Gertman. Therefore, the tube is separate from the true varix in *T. longicornis* as it is in *S. bullisi*. Accordingly, the distinction between *Siphonochelus* s.s. and *Laevityphis* Cossmann seems to have insufficient justification.

FIGURE 32. Muricidae. *Typhis bullisi* Gertman, Sta. P-379, length 11.4 mm.

27. *Typhis* (*Siphonochelus*) *bullisi* (Gertman)
Figs. 32; 34, B; 35, C; 36, C

Siphonochelus (*Laevityphis*) *bullisi* Gertman, 1969, Tulane Stud. Geol., 7 (4): 178, pl. 7, figs. 3a, 3b.

Description.—Gertman, 1969.

Holotype.—U. S. National Museum, No. 696660.

Type-Locality.—OREGON Sta. 5727, southwestern Caribbean Sea, W of Cabo Tiburon, Panama, 8°47'N, 77°09'W, 79 meters, October 17, 1965.

Records.—Seventy-four specimens from 15 stations of R/V PILLSBURY: P-347, Gulf of Uraba (8°43.0'N, 77°03.0'W, 55-53 m, 11 July 1966; 4).—P-361, Colombia (8°51.9'N, 76°37.2'W, 37 m, 12 July 1966; 1).—P-366, Colombia (9°31.0'N, 75°59.5'W, 37-33 m, 13 July 1966; 1).—P-367, Colombia (9°31.1'N, 75°49.6'W, 37-35 m, 13 July 1966; 3).—P-368, Colombia (9°31.2'N, 75°41.1'W, 37 m, 13 July 1966; 21).—P-369, Colombia (9°35.7'N, 75°37.6'W, 18 m, 13 July 1966; 8).—P-370, Colombia (9°37.9'N, 75°50.4'W, 37 m, 13 July 1966; 9).—P-378, Colombia (9°54.6'N, 75°42.4'W, 51-59 m, 14 July 1966; 3).—P-379, Colombia (10°02.2'N, 75°41.3'W, 55 m, 14 July 1966; 1).—P-380, Colombia (10°06.5'N, 75°48.1'W, 64-70 m, 14 July 1966, 1).—P-396, Colombia (9°18.2'N, 76°24.8'W, 70-68 m, 17 July 1966; 2).—P-398, Colombia (9°05.6'N,

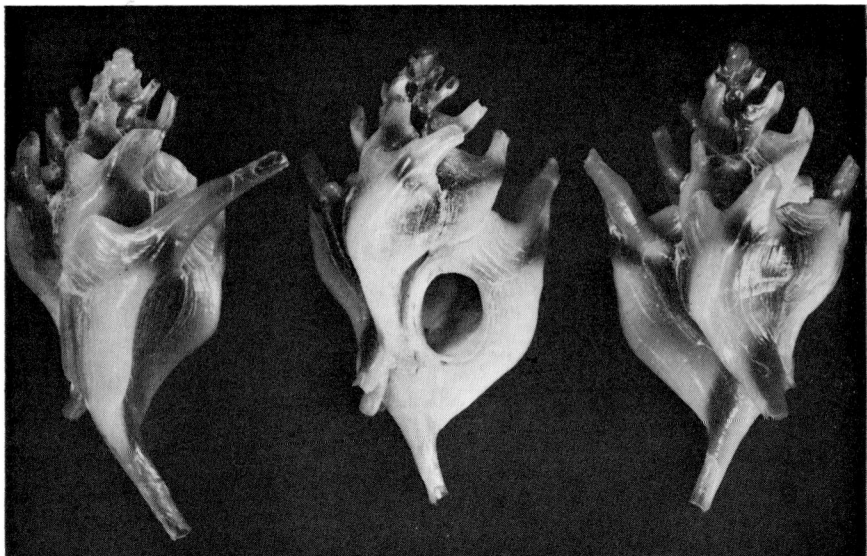

FIGURE 33. Muricidae. *Typhis tityrus*, n. sp., holotype, Sta. P-718, length 11.9 mm.

76°32.1'W, 175-117 m, 17 July 1966; 1).—P-400, Colombia (8°52.4'N, 76°50.4'W, 92-99 m, 17 July 1966; 7).—P-402, Colombia (8°51.2'N, 77°01.6'W, 73 m, 17 July 1966; 7).—P-796, Colombia (10°20.7'N, 75°39.1'W, 60-66 m, 1 August 1968; 5).

Remarks.—These records show *T. bullisi* to be locally abundant in the vicinity of the Gulf of Uraba at depths from 18 to 117 meters.

Although the varical spine is hollow in *T. bullisi* (as Dall noted in *T. linguiferus* also), its lumen does not communicate with the interior of the shell after the spine is fully developed.

28. **Typhis (Siphonochelus) tityrus** n. sp.
Figs. 33; 34, C

Description.—Shell small, composed of 1½ bulbous nuclear whorls followed by five postnuclear whorls. Color rusty brown or pink, the varices white except for a patch of color at the base of the tubes. Interior of aperture flushed with pink. Four smooth, thick, somewhat reflected varices per whorl, without any trace of varical expansion, extending upward to meet the corresponding varix of the preceding whorl. Tubes directed apically, strongly recurved, forming a stout, tapered, hollow spine incorporated in the shoulder of each varix; tips of tubes delicate, glassy, more or less broken

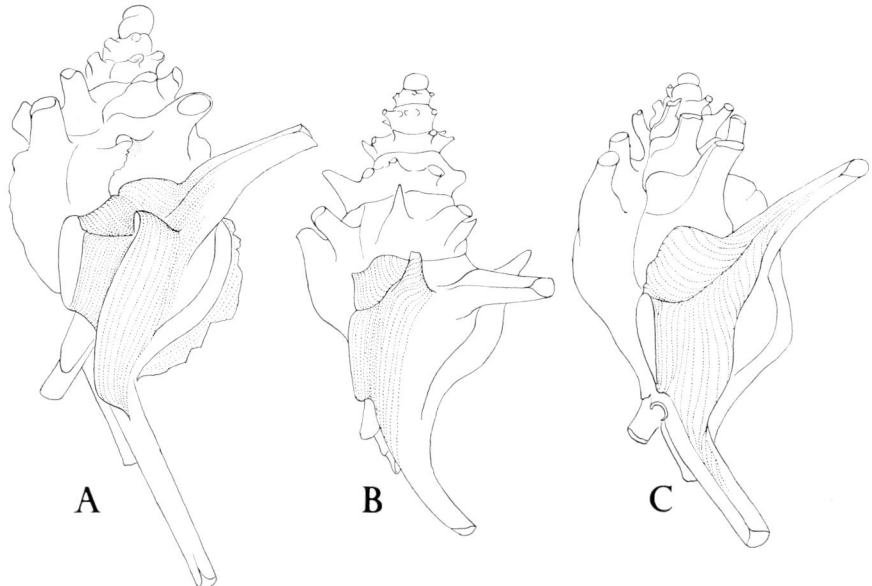

FIGURE 34. Profiles of aperture and final varix: A, *Typhis longicornis* Dall, Sta. G-720; B, *Typhis bullisi* Gertman, Sta. P-379; C, *Typhis tityrus*, n. sp., Sta. P-718.

in all specimens; bases of tubes extended into a thick spiral ridge at the shoulder of the whorls. Suture distinct, impressed. Final varix forming a broad outer lip of nearly uniform width, obtusely angled at the base of the body whorl. Aperture subcircular, its protruding rim externally sculptured with strong growth lines and internally forming three blunt denticles within the palatal margin of old specimens (obscure or absent in younger shells). Anterior canal closed, straight, basally tapered, deflected to the right and abaperturally, long and terminally very delicate and therefore broken to some extent. Surface glossy, with faint, raised spiral lines on the body whorl below the shoulder; incised lines of growth distinct.

Operculum broadly oval, with apical nucleus and strongly raised lamellar lines of growth. Radula not obtained.

Measurements.—Holotype, Sta. P-718: length, 11.9 mm (anterior canal broken), width 6.2 mm.

Holotype.—USNM, No. 700005.

Type-Locality.—Off Isla Margarita, Venezuela, 11°22.5′N, 64°08.6′W, 60 meters, PILLSBURY Sta. P-718, 20 July 1968.

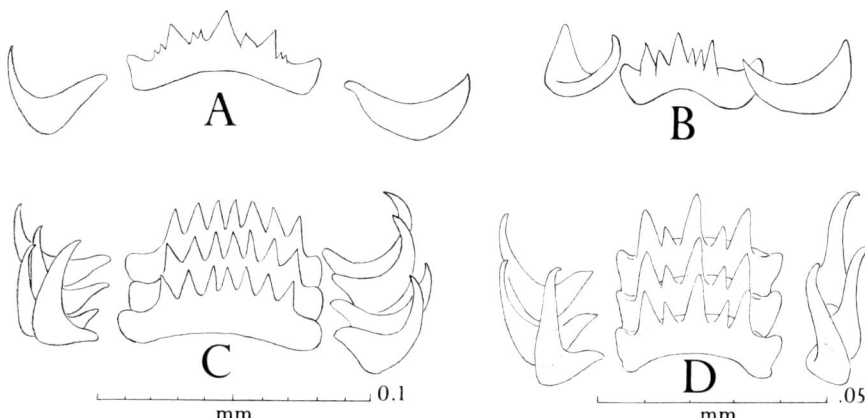

FIGURE 35. Radulae: A, *Typhis longicornis* Dall; B, *Typhis expansus* Sowerby; C, *Typhis bullisi* Gertman; D, *Murex actinophorus* Dall. The 0.1-mm scale applies to A-C; the 0.5-mm scale applies only to D.

Paratypes.—In addition to the holotype, the type-series contains paratypes from the following stations: P-718, type-locality (11°22.5'N, 64°08.6'W, 60 m, 20 July 1968, 1 sp., 8.3 mm × 6.2 mm).—P-727, Gulf of Cariaco (10°20'N, 65°02'W, 64 m, 21 July 1968, 1 sp., 9 mm × 5.2 mm).—P-757, off Pen. de Paraguana (11°39.6'N, 69°22.1'W, 161-187 m, 27 July 1968, 2 sp., 10 mm × 6.5 mm, 9.5 mm × 5 mm).—P-848, N of Trinidad (11°22.0'N, 61°26.4'W, 146 m, 2 July 1969, 1 sp., 13.6 mm × 5 mm).

Remarks.—This small shell bears a strong resemblance to the unique *T. cercadicus* Maury as described by Gertman (1969: 168), but in that species the varices end at the shoulder whereas in the present species they form partitions extending between the shoulder and the suture where they are cemented to the varices of the preceding whorl. In *T. cercadicus* the outer lip narrows above the aperture, whereas in the present form it is somewhat wider.

29. *Typhis (Talityphis) expansus* Sowerby
Fig. 37, A-C

Typhis expansus Sowerby, 1874, Proc. Zool. Soc. Lond. (for 1873): 719, pl. 59, fig. 4.
Typhis (Talityphis) expansus, Gertman, 1969, Tulane Stud. Geol., 7(4): 167, pl. 5, figs. 5a, b; 6a, b (synonymy; designates Paramaribo, Surinam, as type-locality).

Description.—Gertman, 1969: 167.

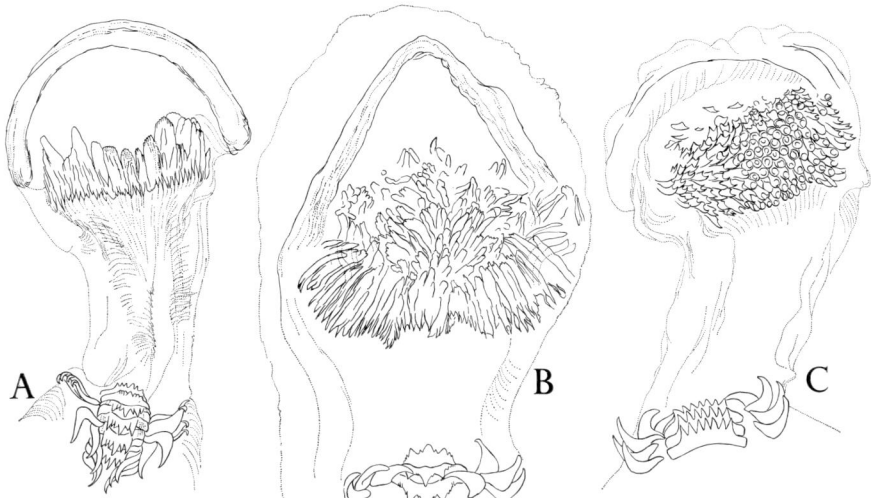

FIGURE 36. Jaws: A, *Typhis expansus* Sowerby; B, *Typhis longicornis* Dall; C, *Typhis bullisi* Gertman.

Material Examined.—Material from the following stations shows definitive formation of the outer lip: P-332, off Punta San Blas (9°31.2'N, 78°53'W, 51 m, 8 July 1966; 2).—P-333, Golfo de San Blas (9°33'N, 78°49'W, 57 m, 8 July 1966; 1).—P-365, Golfo de Morrosquillo (9°31.3'N, 76°15.4'W, 56-58 m, 13 July 1966; 5).—P-402, Golfo de Uraba (8°51.2'N, 77°01.6'W, 73 m, 17 July 1966; 1).—P-574, off Cabo Gracias a Dios (16°16'N, 82°26.5'W, 37 m, 20 May 1967; 1).—P-648, off French Guiana (5°26'N, 52°12'W, 42 m, 8 July 1968; 3).—P-669, off Surinam (6°39'N, 55°15'W, 33 m, 10 July 1968; 1).—P-684, off Surinam (7°19'N, 56°51'W, 55-59 m, 14 July 1968; 1).—P-686, off Surinam (7°00'N, 57°08'W, 27-26 m, 15 July 1968; 5).—P-687, off Guyana (7°13'N, 57°36'W, 27 m, 15 July 1968; 2).—P-699, off Venezuela (9°30'N, 60°15'W, 64 m, 16 July 1968; 1).—P-714, off Margarita I., Venezuela (11°29'N, 64°24.3'W, 59 m, 20 July 1968; 1).—P-721, off Margarita I., Venezuela (11°06.5'N, 64°22.5'W, 26-27 m, 21 July 1968; 3).—P-727, Gulf of Cariaco (10°20'N, 65°02'W, 64 m, 21 July 1968; 2).—P-731, Gulf of Cariaco (10°20'N, 65°41'W, 57-60 m, 22 July 1968; 2).—P-766, off Gulf of Venezuela (12°14.3'N, 70°40'W, 64 m, 28 July 1968; 3).—P-835, off Trinidad (9°36'N, 60°10'W, 48 m, 30 June 1969; 3).

Specimens from the following stations lack the widely expanded, definitive outer lip: P-324, off Punta Manzanillo (9°44'N, 79°31'W, 64-55 m, 7 July 1966; 1).—P-435, Golfo de los Mosquitos (9°08.5'N, 80°29.5'W,

FIGURE 37. Muricidae. A-C, *Typhis expansus* Sowerby; Sta. P-699, length 30.5 mm; D, *Typhis pinnatus* Broderip, Sta. P-368, length 13.5 mm.

37-48 m, 20 July 1966; 1).—P-444, Golfo de los Mosquitos (8°57.5′N, 81°31′W, 73 m, 21 July 1966; 1).—P-619, off Cabo de Honduras (15° 58.2′N, 87°34′W, 18-64 m, 20 March 1968; 1).—P-695, off Guyana (8°12′N, 58°33′W, 37 m, 15 July 1968; 1).—P-842, off Tobago (11°10.6′ N, 60°31.2′W, 68-73 m, 1 July 1969; 2).

Remarks.—The fully developed specimens agree satisfactorily with the description and figures given by Gertman (1969) and no doubt represent Sowerby's species, although they vary greatly in size. Specimens without the expanded lip have a very different aspect and could be taken for a different species. However, individuals showing the formative stage of the definitive lip demonstrate that these forms are merely an ontogenetic stage of *T. expansus.*

30. *Typhis* (*Pterotyphis*) *pinnatus* (Broderip)
Fig. 37, D

Typhis pinnatus Broderip, 1833, Proc. Zool. Soc. Lond., Part 2: 178.
Typhis fordi Pilsbry, 1943, Nautilus, *57*: 40, pl. 7, fig. 4.
Pterotyphis (*Pterotyphis*) *pinnatus*, Gertman, 1969, Tulane Stud. Geol., *7* (4): 183, pl. 8, figs. 2a, 2b.

Description.—Gertman, 1969.

Record.—PILLSBURY Sta. P-368, Golfo de Morrosquillo, Colombia: 9° 31.2′N, 75°41.1′W, 37 meters, 13 July 1966; one specimen.

Remarks.—First reported from Panama by Gertman (1969), this species is now recorded from Colombia.

Family Columbariidae
Genus *Columbarium* von Martens, 1881

Columbarium von Martens, 1881, Conchologische Mittheilungen, *2*: 105.—Thiele, 1929, Handb. Syst. Weichtierkunde, *1*: 289.—Clench, 1944, Johnsonia, *1*(15): 1.—Darragh, 1969, Proc. R. Soc. Vict., *83*: 71.

Diagnosis.—See Thiele, 1929; Clench, 1944; Darragh, 1969.

Type-Species.—*Pleurotoma* (*Columbarium*) *spinicincta* von Martens by original designation (see Darragh, 1969:71).

Remarks.—Until the present time, five species of this genus have been reported in the Western Atlantic: *C. sarissophorum* (Watson, 1882), *C. atlantis* Clench & Aguayo, 1938; *C. bartletti* Clench & Aguayo, 1940; *C. bermudezi* Clench & Aguayo, 1938; and *C. brayi* Clench, 1959. All of these species agree in the elongate, fusiform shape of the shell and distinctly carinated whorls, and, insofar as known (*bermudezi* and *brayi*), have radular characters very close to those reported for the genus (see Peile, 1922: 14, fig. 1; Thiele, 1929: 289, fig. 311). Examination of several small, fusiform gastropods taken in the Straits of Florida revealed nuclear and early postnuclear whorls nearly identical with those of *C. bermudezi*. Study of the radulae of these specimens showed dentition virtually indistinguishable from that of *C. bermudezi* and *C. brayi*. Although none of the shells is strongly carinate, in two the shoulder is angled in all the postnuclear whorls and in the other two it is angled in the early post nuclear whorls but not the later ones. The development of low, curved axial ribs crossed by more or less distinct spiral cords results in a fusinid appearance. Only the radular characters and the appearance of the early whorls show these forms to be distinct from *Fusinus*.

Clench, in 1944, reiterated the prediction that several small species originally described in the genus *Fusus* will prove to belong to *Columbarium* when the type-specimens are reexamined. However, none of the western Atlantic species included in the genus *Fusinus* seem referable to *Columbarium*, even the small forms described by Dall.

Darragh (1969) revised the family and considered its paleontological history. He recognized six genera, one of which he established for a western Atlantic species. As the radular characters are so uniform throughout and the differences in protoconch and adult shell are not great, these taxa are here treated at subgeneric level. Since the three new species here described do not fit any of these taxa well, a new subgenus is proposed for them.

31. *Columbarium* (*Histricosceptrum*) *bartletti* Clench & Aguayo
Fig. 38, A

Columbarium bartletti Clench & Aguayo, 1940, Mem. Soc. Cub. Hist. nat., *14*: 86, pl. 14, fig. 3.—Clench, 1944, Johnsonia, *1*(15): 3, pl. 1, fig. 5.
Histricosceptrum bartletti Darragh, 1969, Proc. R. Soc. Vict., *83*: 88.

Type-Locality.—BLAKE Sta. 9, off Homers Cove, Jamaica: 18°12'N, 78°20'W, 254 fathoms (= 562 m).

Record.—PILLSBURY Sta. P-1225. SW of Jamaica: 17°42.5'N, 78°58.0'W, 546-528 meters, 6 July 1970.

Remarks.—Previously known only from the holotype and one paratype from BLAKE Sta. 9, this species is represented in the PILLSBURY collections by a single example trawled near the type-locality off Jamaica. Although it is a dead shell showing signs of deterioration, it conforms in all respects with the description of the original material.

→

FIGURE 38. Columbariidae. A, *Columbarium bartletti* Clench & Aguayo, Sta. P-1225, length 34.5 mm; B, *Columbarium bermudezi* Clench & Aguayo, Sta. G-1018, length 29 mm (anterior canal incomplete); C, apical whorls of D; D, *Columbarium bermudezi* Clench & Aguayo, Sta. G-1018, length 28.6 mm.

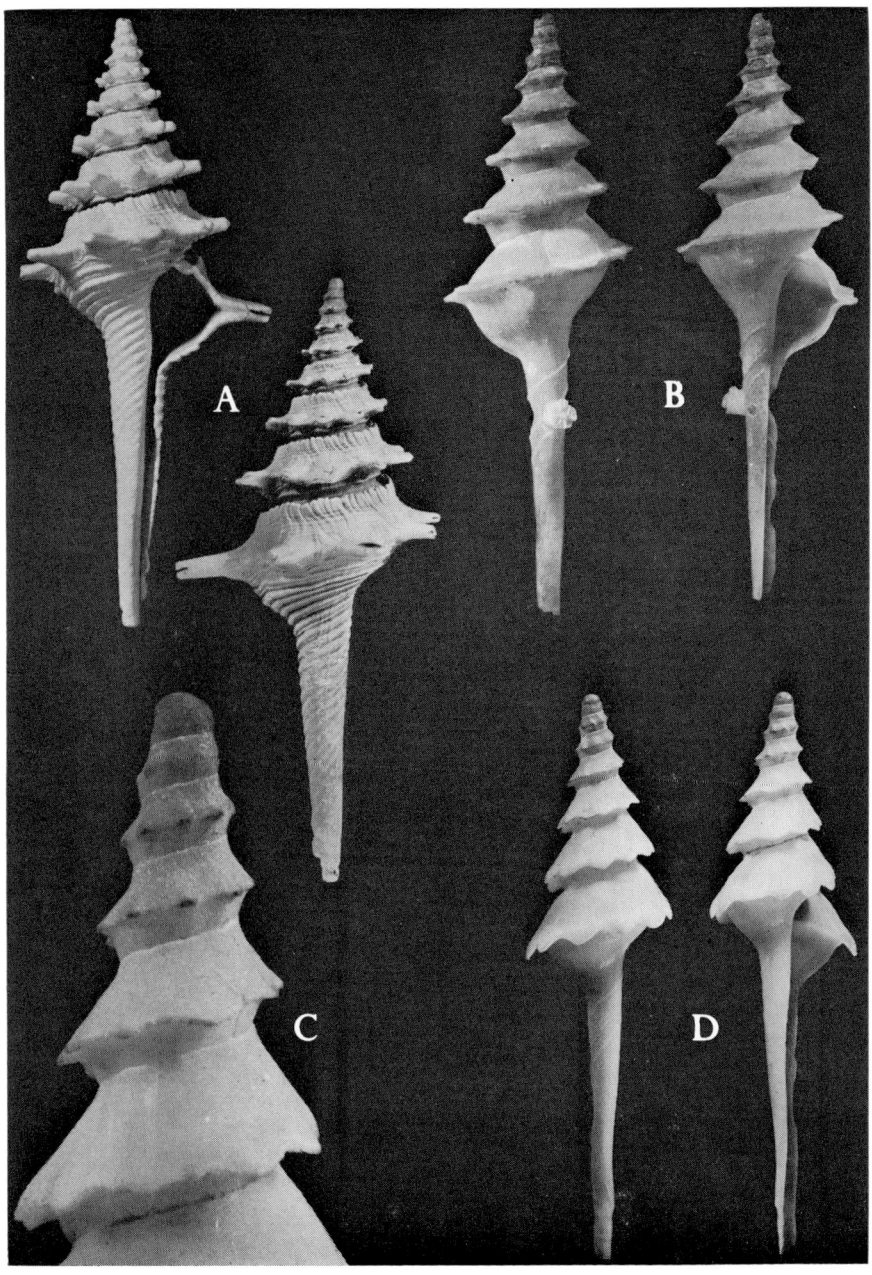

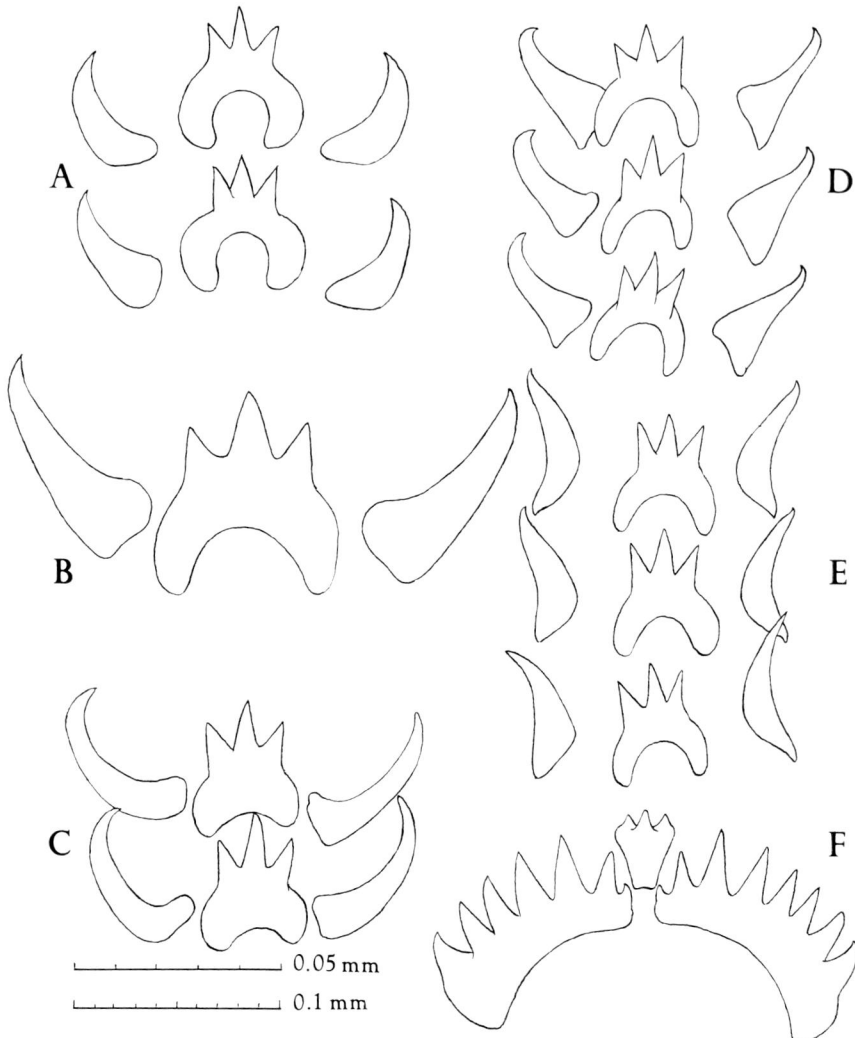

FIGURE 39. Radulae: A, *Columbarium bermudezi* Clench & Aguayo; B, *C. brayi* Clench; C, *Columbarium merope*, n. sp.; D, *Columbarium electra*, n. sp.; E, *Columbarium aurora*, n. sp.; F, *Fusinus eucosmius* Dall. The 0.05-mm scale applies to A-E; the 0.1-mm scale only to F.

32. *Columbarium* (*Fulgurofusus*) *bermudezi* Clench & Aguayo, 1938
Figs. 38, B-D; 39, A

Columbarium bermudezi Clench & Aguayo, 1938, Mem. Soc. Cub. Hist. Nat., *12*: 383, pl. 28, fig. 7.—Clench, 1944, Johnsonia, *1*(15): 2, pl. 1, fig. 3.
Fulgurofusus bermudezi Darragh, 1969, Proc. R. Soc. Vict., *83*: 101, fig. 5, pl. 6, figs. 120-121.

Material Examined.—Eleven specimens from four stations as follows: GERDA Sta. 190, Northwest Providence Channel, NW of Great Stirrup Cay: 25°57'N, 78°07'W, 733-897 meters, 4 July 1963; 1 dead shell.—G-1015, Santaren Channel, west of Anguilla Cays: 23°34'N, 79°17'W, 525-516 m, 15 June 1968; 2 live shells, 1 dead, 1 fragment.—G-1017, Santaren Channel, midway between Cay Sal Bank and Great Bahama Bank: 23°58'N, 79°17'W, 555 m, 15 June 1968; 1 dead shell.—G-1018, Santaren Channel, ENE of Dog Rocks: 24°07'N, 79°28'W, 556 m, 15 June 1968; 1 live, 4 dead shells.

Remarks.—All eleven specimens obviously represent a single species of *Columbarium*. They agree in general with Clench's account (1944) of *C. bermudezi*, although all but one have a more prominent keel than do Clench's specimens. In two examples, the keel stands vertically at the periphery, but in all others it is distinctly deflected downward. It is curious that none of the specimens examined by Clench, from five ATLANTIS stations north of Cuba, showed this feature, but the variations present among the GERDA specimens make it necessary to identify the later as *C. bermudezi*.

In the present specimens, those that were collected alive show a thin periostracum that extends as a fine, sparse fringe of hairs from the edge of the carina. The radula has a tricuspid rachidian and a lateral with a single stout cusp (Fig. 39, A). The operculum has an apical nucleus.

33. *Columbarium* (*Fulgurofusus*) *brayi* Clench
Figs. 39, B; 40

Columbarium brayi Clench, 1959, Johnsonia *3*(39): 330, pl. 173.
Fulgurofusus brayi Darragh, 1969, Proc. R. Soc. Vict., *83*: 102, pl. 6, figs. 113, 115.

Originally described from two specimens dredged off Cabo Codera, Venezuela, in 150 fathoms (= 273 m), this species has now been taken at several additional localities in the southern Caribbean by R/V PILLSBURY. The material now available permits the first description of radula and operculum, and reveals considerable variation in shell characters.

Material Examined.—Thirty-four specimens[2] from the following stations: P-340, Panama, off Punta Mosquito, 9°14'N, 77°46'W, 307-366 meters,

---

[2] Measurements enclosed in square brackets indicate specimens with apex and/or anterior canal, and peripheral carina damaged.

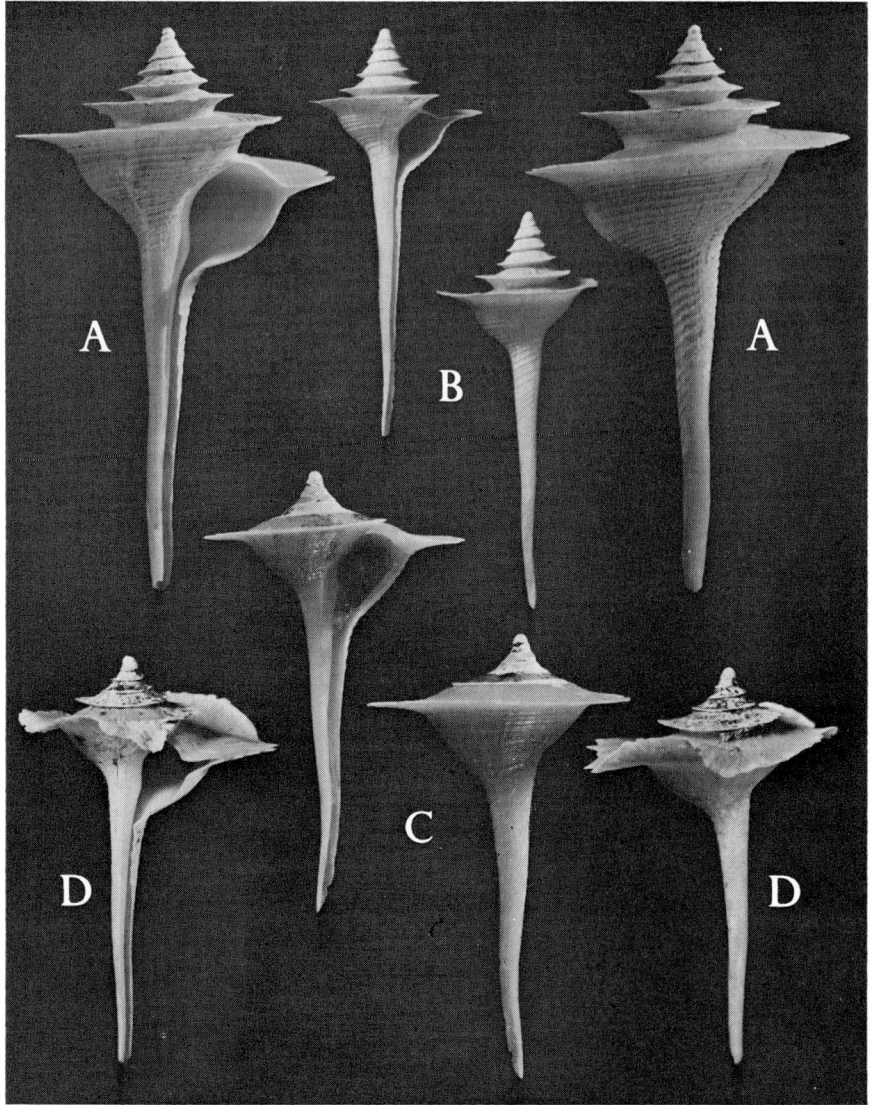

FIGURE 40. Columbariidae. *Columbarium brayi* Clench: A, from Sta. P-374, length 42.8 mm; B, from Sta. P-374, length 31.9 mm; C, from Sta. P-781, length 28.5 mm; D, from Sta. P-445, length 29.6 mm.

9 July 1966; three dead specimens: length 43.9 mm, width [21.6 mm]; length 43.9 mm, width 25.9 mm; length [32.4 mm], width [19.7 mm].—P-374, Colombia, NW of Golfo de Morrosquillo, 9°57′N, 76°11′W, 439-377 meters, 14 July 1966; two dead specimens: length 42.7 mm, width 23.8 mm; length 31.9 mm, width 12.9 mm.—P-386, Colombia, off Cartagena, 10°30′N, 75°42′W, 275-357 meters, 15 July 1966; one dead specimen: length 44.2 mm, width 23.1 mm.—P-394, Colombia, WNW of Golfo de Morrosquillo, 9°29′N, 76°26′W, 421-641 meters, 16 July 1966; eight specimens (6 dead): length [45.9 mm], width [26.4 mm]; length [46.2 mm], width [22.3 mm]; length [32.0 mm], width [20.2 mm]; length 42.1 mm, width 21.3 mm; length [39.0 mm], width [18.4 mm]; length [31.0 mm], width 19.3 mm; length 37.7 mm, width 19.2 mm; length 38.0 mm.—P-445, Panama, ESE of Escudo de Veraguas, 9°02′N, 81°24′W, 342-346 meters, 21 July 1966; six specimens (1 dead): length 35.6 mm, width 21.2 mm; length 29.7 mm, width 18.7 mm; length 29.5 mm, width 19.0 mm; length 28.5 mm, width 17.9 mm; length 27.7 mm, width 19.3 mm; length 27.5 mm, width 16.2 mm.—P-447, Panama, Golfo de los Mosquitos, 9°02′N, 81°07′W, 664-681 meters, 21 July 1966; three specimens (2 dead): length [17.7 mm], width [18.5 mm]; length [22.5 mm], width [18.9 mm]; length [18.0 mm], width [14.9 mm].—P-753, off Venezuela, 11°18.8′N, 68°22.0′W, 384-607 m, 26 July 1968; 1 specimen: length 57.3 mm, width 24.8 mm. —P-776, off eastern Colombia, 12°13.3′N, 72°50′W, 408-576 meters, 29 July 1968; one specimen: length [25.2 mm], width [16.5 mm].—P-781, off eastern Colombia, 11°30.1′N, 73°26.5′W, 567-531 meters, 30 July 1968; nine specimens (3 dead): length [36.3 mm], width 23.2 mm; length [35.9 mm], width 23.1 mm; length [34.8 mm], width [21.5 mm]; length [27.8 mm], width [28.9 mm]; length 28.6 mm, width 16.6 mm; length 26.3 mm, width 15.6 mm; length 17.6 mm, width [11.5 mm]; length 15.1 mm, width [8.4 mm]; length [15.1 mm], width [10 mm].

Operculum.—Thin, yellowish, with terminal nucleus.

Radula.—Rachidian tricuspid; lateral with a single strong, curved cusp (Fig. 39, B).

Variation.—Most of the specimens taken by R/V PILLSBURY differ from the type-material in having a lower spire and the carina placed higher on the whorls, closer to the suture. In the specimens from stations P-445, P-776, and P-781, the spire is so depressed and the suture so close to the carina of the preceding whorl that the typical "pagoda-like" profile is obliterated.

Clench (1959) noted a thin periostracum with short axial blades. In our material, this is so thin that it is almost invisible and there are no detectable "short axial blades" even in the young specimens that were alive when collected.

Range.—Southern part of the Caribbean Sea: the westernmost record is PILLSBURY station P-445 in Golfo de los Mosquitos, Panama (9°02′N, 81°24′W); the easternmost record is the type-locality, off Cabo Codera, Venezuela (11°N, 66°01′W). The species now has been taken from nine localities off Panama, Colombia, and Venezuela. The shallowest record is the type lot, from 150 fathoms (= 273 meters); the deepest verifiable record is 664 meters.

Columbarium brayi appears to be fairly common on mud bottom in the southern Caribbean at depths from 300 to 500 meters.

Peristarium, new subgenus

Diagnosis.—Elongate fusiform shells of small or moderate size, with extended spire and long, straight, tapered siphonal canal. Nucleus bulbous, glassy, of about two whorls, followed by about eight sculptured postnuclear whorls. The first few postnuclear whorls with angular peripheral nodes on low axial ribs, becoming obscure on the subsequent whorls, which are not carinate or spinose. Spiral sculpture consisting of low, narrow cords, more or less distinct. Radula typical of *Columbarium,* with an arched rachidian plate bearing three cusps, flanked on each side by a single flattened, curved, triangular lateral. Operculum thin, corneous, translucent yellow, ovate-unguiculate, with apical nucleus.

Type-Species.—*Columbarium* (*Peristarium*) *electra*, new species; here designated.

Gender.—Neuter.

Remarks.—These shells are similar to fusiform species of *Coluzea* from South Africa. They differ from Darragh's definition of that group by having no sharp distinction between nuclear and postnuclear whorls, and from most other species by lacking a flange-like or dentate keel.

34. Columbarium (Peristarium) electra, new species
Figs. 39, D; 41

Material Examined.—GERDA Sta. G-289. Straits of Florida SSE of Key West: 24°11′N, 81°36′W, 604-594 meters, 3 April 1964. Six specimens.

Description.—Shell elongate fusiform, with elevated spire and long, narrow, tapered siphonal canal. Color white. Whorls nine, regularly increasing in size; nuclear whorls about 1½, bulbous, smooth and glossy, not clearly delimited from the postnuclear whorls except by the initiation of a series of obscure peripheral nodes which become more prominent as the whorls increase in size. Each node is situated on a low, rounded axial ridge; the ridges are at first weak but become stronger on the second and third postnuclear whorls, then gradually decrease in prominence until they are merely

FIGURE 41. Columbariidae. *Columbarium (Peristarium) electra*, n. subgen., n. sp., holotype, Sta. G-289, length 26.85 mm.

low, broad axial undulations, 14 in number on the body whorl. Peripheral nodes low and blunt on the later whorls, where they lie about midway between sutures; 14 nodes occur on the body whorl, one on each axial rib. Growth lines visible on the last four postnuclear whorls. Faint spiral lines begin on about the fourth postnuclear whorl, becoming more distinct on the following whorls where they appear as narrow bands of slightly different texture only slightly raised from the adjacent shell surface. On the spire there are three of these major lines above the periphery, separated by much fainter ones. Three spiral lines follow the periphery and are more distinctly raised where they pass over the peripheral nodes. Below the nodes there is one principal spiral; on the body whorl there are five distinct spirals below the periphery (the suture follows the second) and another eight on the siphonal canal, fading out toward its end. Columella long, straight, imperforate, not plicated. Interior surface of outer wall smooth. Parietal wall without a distinct lip, smooth, the spiral sculpture obliterated. The operculum is ovate-unguiculate, with apical nucleus. The radula has an arched tricuspid rachidian plate flanked by a triangular, clawlike lateral on each side (Fig. 39, D).

Measurements.—Length 26.85 mm, diameter at periphery 6.45 mm (holotype).

Holotype.—USNM No. 701151, from GERDA Sta. G-289.

Type-Locality.—Straits of Florida, SSE of Key West: 24°11′N, 81°36′W, 604-594 m.

Remarks.—The operculum is like that of *C. (Peristarium) merope*, n. sp., and other Caribbean species of the genus. (See Figure 43.)

35. **Columbarium (Peristarium) merope**, new species
Figs. 39, C; 42; 43, A

Material Examined.—GERDA Sta. G-439, Straits of Florida, SW of Marquesas Keys (24°14′N, 82°29′W, 584-566 meters, 29 November 1964; two specimens, paratypes).—G-440, Straits of Florida, SW of Marquesas Keys (24°14′N, 82°21′W, 549-567 meters, 29 November 1964; one specimen, paratype).—G-476, Straits of Florida, SW of Marquesas Keys (24°14′N, 82°24′W, 549-512 meters, 26 January 1965; one specimen, holotype).—G-966, Straits of Florida, SW of Marquesas Keys (24°10′N, 82°22′W, 553-558 meters, 2 February 1968; one specimen, paratype).—G-970, Straits of Florida, S of Marquesas Keys (24°24′N, 82°08′W, 512 meters, 2 February 1968; one specimen, paratype).

Description.—Shell elongate fusiform, with elevated spire and long, narrow, tapering siphonal canal. Whorls 10½, regularly increasing in size; nuclear whorls about two, bulbous, smooth and glossy, not clearly distinguished from the postnuclear whorls. First four whorls closely resembling those of *Columbarium electra*. In the course of the fourth turn, three spiral cords become conspicuous, one at the periphery and two below it. About the sixth whorl, five spirals become quite distinct above the periphery, and an intermediate spiral appears between the peripheral one and that below it, so that three broad spirals follow the periphery and pass over the nodes. Between the three peripheral spirals and the suture below there is at first one spiral cord, then by intercalation of one above and below it there are eventually three. On the body whorl of the largest specimen, there are ten strong spirals below the three peripheral ones, and about 24 extending down the siphonal canal and fading out terminally. Close-set axial growth lines are distinct on all the later whorls. The peripheral nodes occupy the summit of low, rounded, semilunate axial ribs, of which there are 12 on the body whorl of the largest examples. Columella long and straight, with-

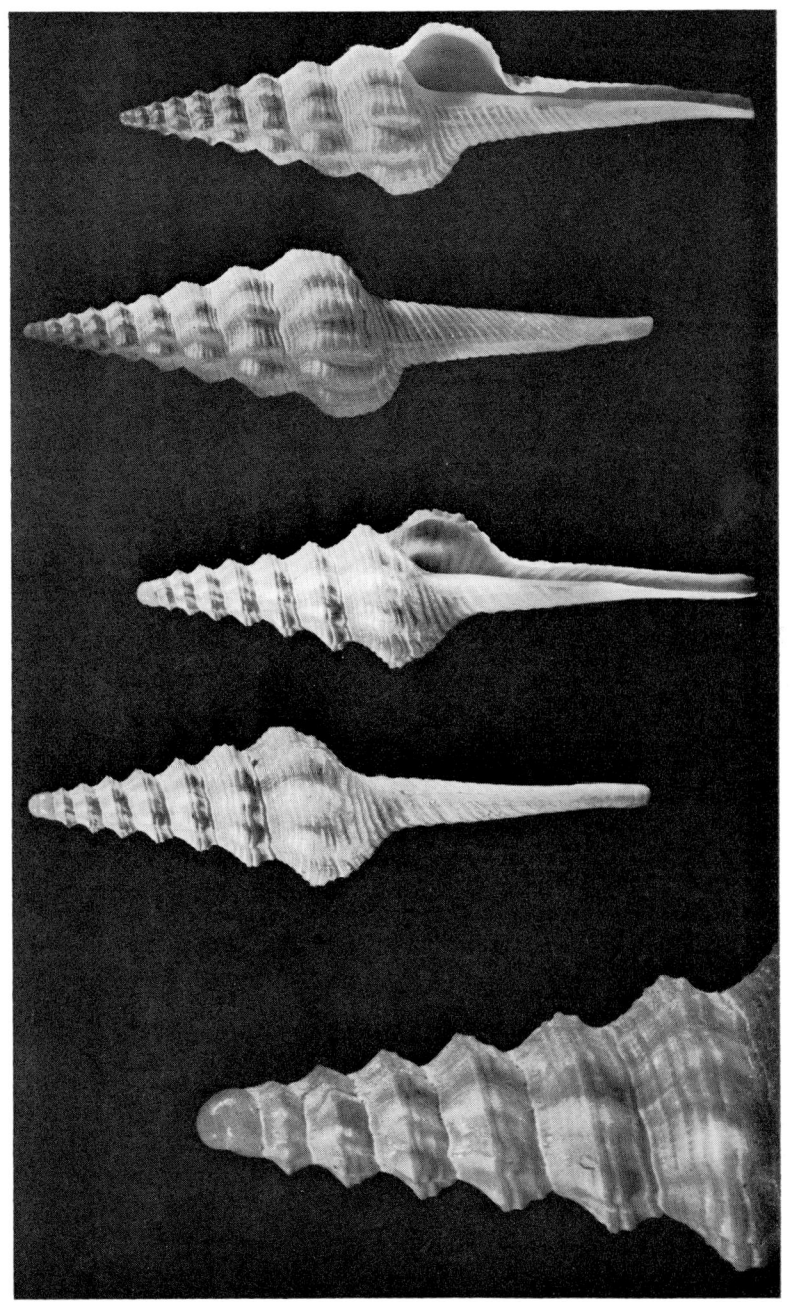

FIGURE 42. Columbariidae. *Columbarium (Peristarium) merope*, n. sp. Left to right: apical whorls of paratype from Sta. G-966; two views of entire shell of same, length 22.4 mm; two views of holotype from Sta. G-476, length 36 mm.

out trace of plications. Interior of aperture shows impressions of the exterior spiral cords in the younger specimens, but in the largest examples the interior of the outer wall is quite smooth. Parietal wall smooth, due to the removal of the sculptured superficial layer of shell. Periostracum thin, faintly yellowish, projecting along the growth lines as thin, narrow lamellae which are elongated into small, pointed projections along the raised spiral cords.

The operculum (Fig. 43, A) is like that of other species, yellowish, translucent, with apical nucleus. A colony of hydroids is attached to its outer surface in several cases.

The radula is virtually indistinguishable from that of other species of *Columbarium* (Fig. 39, C).

Measurements.—Length 36.0 mm, diameter at periphery 9.7 mm (holotype).

Holotype.—USNM No. 701152, from GERDA Sta. G-476.

Type-Locality.—Southern part of the Straits of Florida, SW of Marquesas Keys, 24°14'N, 82°24'W, 540-512 meters.

Remarks.—This species is distinguished from *Columbarium (Peristarium) electra* by its stronger spiral sculpture. From *C. (P.) aurora*, which is similarly sculptured, it is distinguished by its more angular whorls, by the more abrupt transition of the body whorl basally into the siphonal canal, and by the proportionally higher spire and shorter siphonal canal.

All of the known specimens were taken in the lower Straits of Florida in an area south of the Marquesas Keys in depths between 512 and 584 meters. *Columbarium aurora* so far has been taken only from the straits in the area off Miami, Florida. Many stations made at comparable depths in intermediate localities have not yielded specimens of either species.

36. **Columbarium (Peristarium) aurora**, new species
Figs. 39, E; 43, B; 44

Material Examined.—GERDA Sta. G-62, Straits of Florida, SE of Fowey Light (25°30.5'N, 80°00'W, 403-384 m, 29 August 1962; one specimen, paratype).—G-66, Straits of Florida, SE of Fowey Light (25°25.5'N, 79°59'W, 366 m, 26 September 1962; one specimen, paratype).—G-67, Straits of Florida, SE of Fowey Light (25°31'N, 79°57'W, 351 m, 26 September 1962; two specimens, paratypes).—G-828, Straits of Florida, E of Fowey Light (25°34'N, 79°57'W, 333-340 m, 7 July 1967; one specimen, paratype).—PILLSBURY Sta. P-1309, Straits of Florida, NE of Fowey Light (25°40'N, 80°02'W, 247 m, 5 December 1970; four specimens, holotype and paratypes).

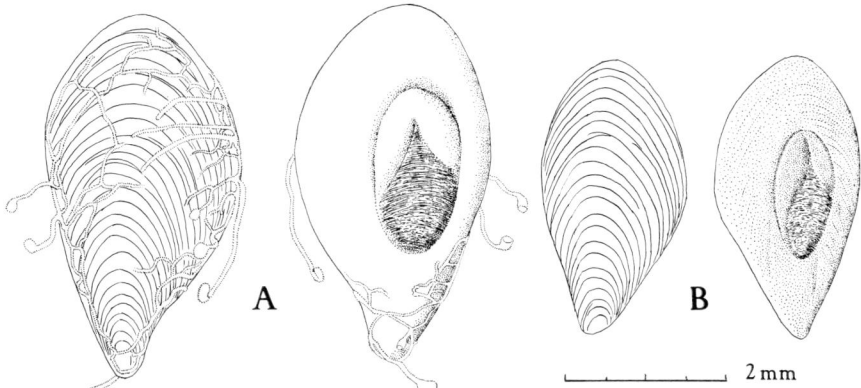

FIGURE 43. Opercula: A, *Columbarium (Peristarium) merope*, n. sp.; B, *Columbarium aurora*, n. sp.

Description.—Shell elongate fusiform, with elevated spire and prolonged siphonal canal much as in *Columbarium merope*. Color white, some specimens more or less strongly suffused with pink, especially on the body whorl and siphonal canal. Whorls ten, of which two form the smooth, bulbous nucleus. After the second nuclear whorl, low axial ribs with a distinct angular projection at the periphery appear and become stronger on the following whorls. Third whorl with about 11 axials, body whorl with 12-16. About the third turn, two spiral cords appear above the periphery; one passes over the peripheral nodes, and one below the periphery close to the suture. By the sixth whorl there are about six spirals above the peripheral one (including the broad subsutural one) and from one to three below it; on the body whorl, there are 16-23 spirals followed on the siphonal canal by as many as 40 (of which several are narrow intercalaries); those near the tip of the canal become weak but remain visible up to the line along which the direction of the growth lines changes near the end of the canal. On the early whorls, the spirals are narrow, flat, separated by interspaces wider than the cords. On the later whorls the cords become progressively wider until, on the body whorl, they are wider than the interspaces which, in some cases, are reduced to narrow grooves. Although the axial sculpture persists on all the postnuclear whorls, the peripheral angulation becomes less distinct from the sixth whorl onward and is hardly perceptible on the body whorl. The columella is long, nearly straight, lacking plications. The tip of the siphonal canal is more or less deflected upward, especially in larger specimens. Inside of outer lip smooth; parietal wall smooth where the mantle has removed the sculpture, without raised

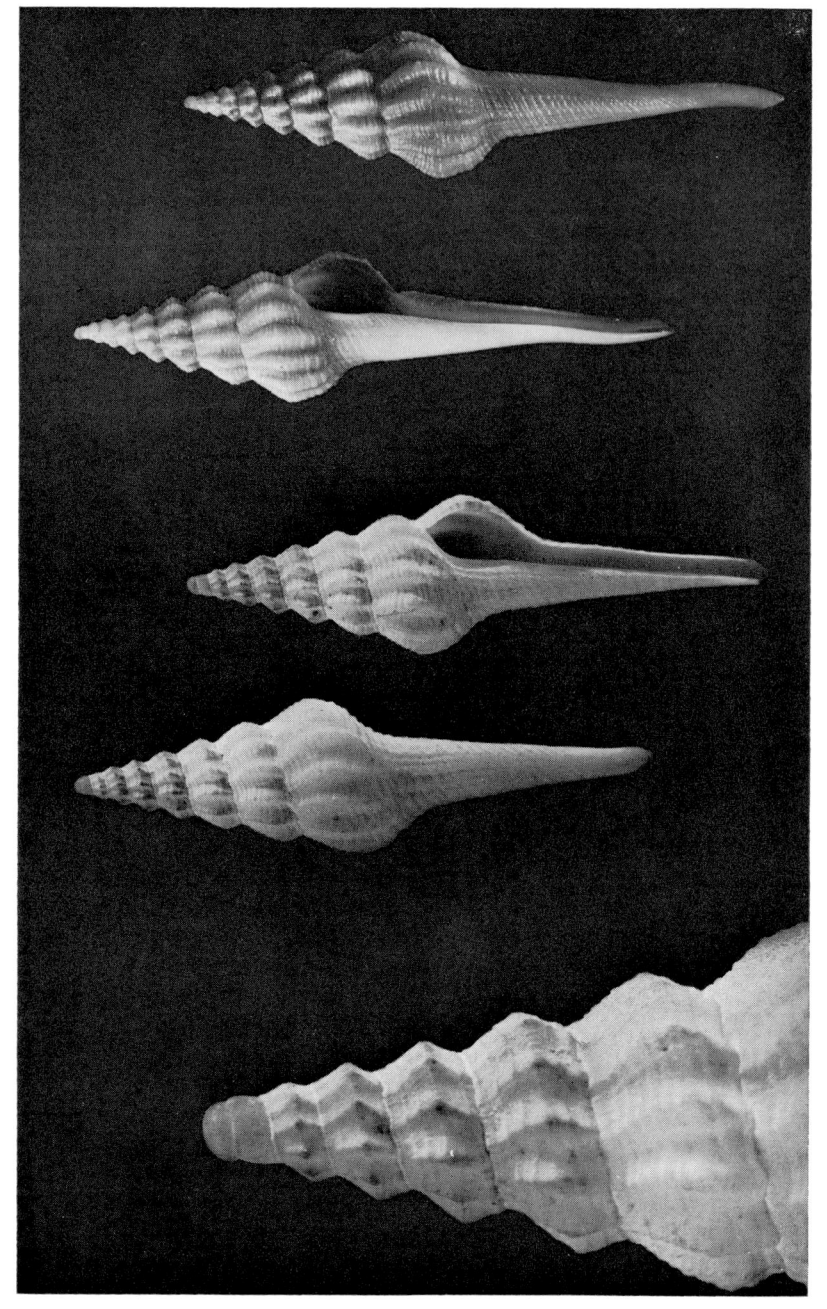

FIGURE 44. Columbariidae. *Columbarium aurora*, n. sp. Left to right: apical whorls of paratype from Sta. G-62; two views of entire shell of same, length 25.9 mm; two views of holotype from Sta. P-1309, length 41.4 mm.

parietal lip. Periostracum thin, yellowish, forming short projecting fringes along the course of the spiral cords, most noticeable in small specimens.

The operculum (Fig. 43, B) is like that of the other species herein described. The radula (Fig. 39, E) conforms closely to that of other species of *Columbarium*.

Measurements.—Length 41.4 mm, diameter at periphery 10.5 mm (holotype); length 25.9 mm, diameter 7.2 mm (paratype); largest specimen, length 43 mm (apex decollated), diameter 11.3 mm (P-1309); smallest specimen, length 22.4 mm, diameter 6.1 mm (G-67).

Holotype.—USNM No. 701222, from PILLSBURY Sta. P-1309.

Paratypes.—USNM No. 701153, from GERDA Sta. G-62. Others in the collection of the Rosenstiel School of Marine and Atmospheric Sciences.

Type-Locality.—Straits of Florida, off Fowey Light, 25°40'N, 80°02.8'W, 247 m.

Remarks.—This species is distinguished from *Columbarium* (*Peristarium*) *merope* by the rounded shoulder of the later whorls, the smoother gradation of the body whorl into the siphonal canal, which is more conspicuously tapered, and the relatively shorter spire and longer siphonal canal.

Columbarium aurora, like *C. electra* and *C. merope*, closely resembles a small *Fusinus*. All are distinguished from that genus by both radular and nuclear characters. Figures of the shell and apex (Fig. 45) and of the radula (Fig. 39, F) of *Fusinus eucosmius* Dall are introduced here for comparison.

Family Coralliophilidae
(= Rapidae, = Magilidae)

The genus *Coralliophila* has been subdivided to such an extent that it is difficult, if not impossible, to assign many species unequivocally to a genus-group taxon. *Latiaxis* Swainson, established for *Pyrula mawae* Griffith & Pidgeon, characterized by its depressed spire, imbricated flattened spines at the shoulder and margin of the umbilicus, and loosely coiled whorls with the spiral sculpture otherwise not strongly imbricated, is distinctive but should be restricted to species such as *L. mawae* and *L. pilsbryi* Hirase. No species of this group has been found in the West Atlantic. The allocation of species such as *C. deburghiae* (from Japan) and *C. dalli* (from Florida), with elevated spire as well as prominent flat spines at the shoulder, to the genus *Latiaxis* seems to lack justification. Accordingly, I prefer to assign all the western Atlantic forms to *Coralliophila* sensu lato. Except for *C. lactuca* Dall, with its strong axial sculpture, there is no reliable generic distinction among them.

FIGURE 45. Fusinidae. *Fusinus eucosmius* Dall, Sta. G-1039, length 55 mm; shell (periostracum removed) and apical whorls.

Genus *Coralliophila* H. & A. Adams

Coralliophila H. & A. Adams, 1853, Gen. Recent Mollusks, *1*: 135.—Dall, 1889a, Bull. Mus. comp. Zool. Harv., *18*: 217.—Abbott, 1958, Monogr. Acad. nat. Sci. Philad., no. 11: 65.

Latiaxis (*partim*), Auctt., *non* Swainson, 1840.

Abbott (1958) has reviewed the shallow-water species of this genus known from the Caribbean area. Some of them descend into moderate depths and may be taken by dredging.

37. *Coralliophila dalli* (Emerson & D'Attilio)
Fig. 46

Coralliophila deburghiae, Dall, 1889a, Bull. Mus. comp. Zool. Harv., *18*: 218, pl. 16, fig. 5; 1889b, Bull. U. S. natn. Mus., *37*: 122, pl. 16, fig. 5 (also second ed. in 1903).—Smith, 1937, East Coast Marine Shells: 118

FIGURE 46. Coralliophilidae. *Coralliophila dalli* (Emerson & d'Attilio): upper, three views of specimen from Sta. G-1206, height 28 mm, width overall, 31.5 mm; lower, two views of specimen from Sta. G-394, height 33.1 mm, width overall 24 mm.

(but not pl. 58, fig. 5); 1939, Illustr. Cat. Rock Shells: 32, in part (pl. 20, fig. 11?).

Not *Pyrula* (*Rhizochilus*) *deburghiae* Reeve, 1857, Proc. Zool. Soc. Lond. Part 25: 208, pl. 38, fig. 3a, b.

Latiaxis (*Babelomurex*) *dalli* Emerson & D'Attilio, 1963, Am. Mus. Novi., No. 2149: 4, figs. 1-3.

Material Examined.—GERDA Sta. G-394, NW of Little Bahama Bank (27°22'N, 79°11'W, 223 meters, 19 September 1964; 2).—G-1206, North-

FIGURE 47. Coralliophilidae. *Coralliophila mansfieldi* (McGinty), Sta. P-405, height 18.3 mm.

west Providence Channel (26°17'N, 78°54'W, 394-420 meters, 20 August 1969; 1).

PILLSBURY Sta. P-890, Lesser Antilles, NE of St. Lucia (14°05.6'N, 60°51.4'W, 198-430 meters, 7 July 1969; 1).—P-943, Lesser Antilles, off Grande Terre, Guadeloupe (16°25.9'N, 61°36.7'W, 275 meters, 17 July 1969; 8).

Remarks.—This long-misidentified species, first figured by Dall (1889a), has been obtained at several localities during explorations by R/V GERDA and R/V PILLSBURY. Ornate examples of *Coralliophila mansfieldi* from shallow water sometimes have been called "*Coralliophila deburghiae*" in error; the specimen figured by Smith (1939, pl. 20, fig. 11) may be such a shell.

Although the flattened, up-curved peripheral blades vary as to length and degree of curvature, they appear to be characteristic of the species. Specimens with the spines suppressed, treated by Dall under the varietal names *fusiformis* and *Lintoni*, are here considered specifically distinct.

In *Coralliophila dalli*, the spiral sculpture is much more prominent and more strongly imbricated than in *C. fax* and *C. sentix*, new species. On the upper slope of the whorls, one of the spiral cords (third or fourth from the suture) is more prominent and lies about midway between the suture and the base of the carina.

38. *Coralliophila mansfieldi* (McGinty)
Fig. 47

Muricidea mansfieldi McGinty, 1940, Nautilus, 53: 83, pl. 10, figs. 5, 5a. (Fossil, Caloosahatchee Marl, Clewiston, Florida.)
Coralliophila mansfieldi, Pilsbry & McGinty, 1949, Nautilus, *63*: 11, pl. 1, figs. 2-3. (Recent, Palm Beach Inlet, Biscayne Bay, and Destin, Florida.)
Latiaxis (Babelomurex) mansfieldi, Emerson & D'Attilio, 1963, Am. Mus. Novit., No. 2149: 7, fig. 4.

This species, not uncommonly taken in shallow water in southern Florida from Palm Beach Inlet southward, sometimes was identified as *C. deburghiae* by collectors. It was tentatively synonymized with *C. scalariformis* (Lamarck) by Abbott (1958: 66), but it does not very closely resemble that species as figured by Kiener (1836: pl. 19, fig. 55). Specimens obtained by R/V PILLSBURY in the southwestern Caribbean (Fig. 47) do not differ in any significant way from shallow-water specimens from Florida.

Record.—PILLSBURY Sta. P-405, Gulf of Uraba, 8°49'N, 77°21'W, 92-93 m, 17 July 1966. Two specimens.

39. *Coralliophila tectumsinensis* (Deshayes)
Fig. 48

Murex Tectum Sinense Deshayes, 1856, J. Conchyliol., *5*: 78, pl. 3, figs. 1-2 (Habite les côtes de l' Algérie).
Rhizochilus (Coralliophila) bracteata, part, Tryon, 1880, Man. Conch., *2*: 210, pl. 66, fig. 380.

Material Examined.—PILLSBURY Sta. P-650, off French Guiana (6°07'N, 52°19'W, 46-50 meters, 8 July 1968; 1).—P-729, off Venezuela, SE of Isla Tortuga (10°41.1'N, 65°17.4'W, 720-725 meters, 22 July 1968; 1).

Description.—Shell of rather small size and solid structure, with elevated spire, the whorls strongly shouldered, carinated; body whorl somewhat inflated, anterior canal short but distinct, siphonal fasciole prominent, enclosing a shallow, funnel-like umbilical excavation. Aperture broadly ovate, narrowing toward the siphonal canal. First nuclear whorl lost, total number of whorls probably eight. Sculpture consisting of low, rounded axial ribs beginning about the third whorl and numbering about ten on the body whorl, crossed by imbricated spiral cords. Shoulder carinated, ornamented with flattened, triangular, upturned spines where it is crossed by the axial ribs. Slope above the carina sculptured by about ten spiral cords with thin, raised, imbricating scales; upper surface of carina with about three smaller,

FIGURE 48. Coralliophilidae. *Coralliophila tectumsinensis* (Deshayes) Sta. P-729, height 24 mm.

scaled cords extending out onto the spines. On the body whorl there are about seven major cords below the carina, separated by another six or seven secondary cords, all conspicuously imbricated where crossed by the axial growth-lamellae. Interior of outer lip with nine smooth, sharp lirations; parietal wall smooth, obliterating the surface sculpture, somewhat raised at its margin to form a thin parietal lip. Color white, the upper whorls tinged with light brown; aperture white.

Measurements.—Length 12.5 mm, width overall 10 mm (P-650); length 24 mm, width overall 17.1 mm (P-729).

Remarks.—A specimen of *Coralliophila* from the Weinkauff collection of Mediterranean shells now in the U. S. National Museum of Natural History, originally identified as *C. tectumsinensis* Deshayes, agrees closely with the present material, and both correspond well with Deshayes' description and figure (1856) of the original specimen from Algeria. *C. babelis* Requien (type-species of the subgenus *Babelomurex* Coen), synonymized with *C. bracteata* Brocchi by Tryon (1880), has a more slender shell with longer anterior canal, and the upper slope of the whorls has only three or four strong, imbricated spirals rather than 18 or 19 narrow ones.

It is possible that the West Indian material will merit subspecific recognition upon more intensive study, as the sculpture is somewhat coarser, especially on the upper slope, than in the Mediterranean specimen in the National Museum collections. However, these shells show so much variation that separating them at this time is not warranted.

40. Coralliophila sentix, new species
Fig. 49

Material Examined.—PILLSBURY Sta. P-876, Lesser Antilles, E of St. Vincent (13°13.9'N, 61°04.7'W, 231-258 meters, 6 July 1969. Two live specimens).—P-903, Lesser Antilles, SW of St. Lucia (13°44'N, 61°03.1' W, 231-430 meters, 9 July 1969; one shell dead but fresh).

Description.—Shell of moderate size and solid structure, with elevated spire, prominent anterior canal, and strongly carinate, shouldered whorls. Columella straight, openly perforate, umbilicus bounded by a distinct siphonal fasciole imbricated by the preceding temporary ends of the siphonal canal. Aperture ovate, the outer lip of mature specimens with about eight internal lirations. Nuclear whorls lost in all specimens examined; largest specimens with seven remaining whorls, total probably nine. Carina moderately wide, rather thick, strongly upturned, irregularly serrated marginally; the marginal projections are extensively damaged in the two larger examples, less so in the smallest one. Projections flat, triangular, with the apex curved backward more or less, sometimes joining the forward edge of the preceding spine. Early postnuclear whorls developing flattened spines where axial ribs cross the shoulder, axial component becoming obscure with increasing size of shell, persisting only as a succession of slightly thickened former outer-lip edges; intermediate axial growth lines inconspicuous. Spiral sculpture distinct, only moderately imbricated. Above the carina the spirals are weaker than below it, the primary ones further marked by very fine secondary spirals, all diverging outward onto the carinal spines. Below the carina there are about 20 spiral cords on the body whorl; another 15 or more, less regular, mark the lower surface of the carina and diverge centrifugally out onto the spines. Color white, more or less flushed with pinkish tan; interior of aperture white, with or without pinkish tints in the throat.

Measurements.—Length 28.6 mm (holotype); length 16.35 mm (paratype, P-876); length 36.2 mm (paratype, P-903).

Holotype.—USNM No. 701155, from PILLSBURY Sta. P-876.

Type-Locality.—East of St. Vincent, Lesser Antilles, Sta. P-876.

Paratype.—USNM No. 701156, from PILLSBURY Sta. P-903.

Remarks.—In general appearance, this species resembles "*Latiaxis*" na-

FIGURE 49. Coralliophilidae. *Coralliophila sentix*, n. sp.: upper, two views of paratype from Sta. P-903, height 36.2 mm; lower, three views of holotype from Sta. P-876, height 28.7 mm.

FIGURE 50. Coralliophilidae. *Coralliophila fax*, n. sp., holotype, Sta. G-1125, height 24.4 mm.

kamigawai Kuroda from Japan (see Shikama & Horikoshi, 1963: pl. 61, fig. 2). It is quite distinct from other West Indian coralliophilids. The broader and more confluent carinal blades and much weaker spiral sculpture distinguish *Coralliophila sentix* from *C. dalli*, whereas it is heavier, coarser, and more strongly sculptured than *C. fax*.

41. **Coralliophila fax**, new species
Fig. 50

Material Examined.—GERDA Sta. 1125, Straits of Florida, NW of Settlement Point, Grand Bahama: 26°45′N, 79°05′W, 494-530 meters, 13 June 1969, one specimen.

Description.—Shell of moderate size, thin and light in construction, with a narrow, elevated spire, moderately produced siphonal canal, and strongly carinate, shouldered whorls. Columella straight, narrowly perforate, umbilicus bounded by a narrow siphonal fasciole marked by imbricated sculpture. Aperture triangular, outer lip without interior lirations. Nuclear whorls missing and early postnuclear whorls eroded; almost 7 whorls remaining. Carina broad, thin, widely upturned, broadly serrated marginally;

the marginal serrations, which are not filled with shelly material, are mostly broken in this specimen, and probably would be more or less so, even in living specimens. The sculpture of the early whorls consists of flattened, probably upturned, spines where low axial ribs cross the shoulder. The third and fourth preserved postnuclear whorls show a distinct raised spiral ridge just above the suture, but this becomes less prominent on the later whorls and is evident on the body whorl only as a slightly more prominent, low spiral cord. Axial growth lines rather irregular but distinct. Spiral sculpture of low, broad threads without prominent imbrication. Shell dead, discolored a uniform dull grey.

Measurements.—Length 24.4 mm, width including carina (edge imperfect) 22.7 mm (holotype).

Holotype.—USNM No. 701154, from GERDA Sta. G-1125.

Type-Locality.—Straits of Florida, off Settlement Point, Grand Bahama, Sta. G-1125.

Remarks.—The broad, marginally serrated carina of this species superficially resembles that of the Japanese *Latiaxis mawae* (Griffith & Pidgeon), but in that species the spire is depressed, the parietal lip is free of the preceding whorl in mature shells, and the umbilicus is widely open. From *Coralliophila dalli*, this species differs in the form of the carina, the nearly smooth surface, and the narrow umbilical opening.

42. *Coralliophila lamellosa* (Philippi)
Fig. 51

Fusus lamellosus Philippi, 1836, Enumeratio Molluscorum Siciliae, *1*: pl. 11, fig. 30.
Trophon Lintoni Verrill & Smith, 1882, Amer. Journ. Sci., *24*: 365.—Verrill, 1884, Trans. Conn. Acad., 6: 176, pl. 29, fig. 1.
Coralliophila Deburghiae Reeve var. *Lintoni*, Dall, 1889a, Bull. Mus. comp. Harv., *18*: 219; 1889b, Bull. U. S. Nat. Mus., *37*: 122, 190, pl. 44, fig. 1.—Smith, 1939, Illustr. Cat. Rock Shells: 32, pl. 15, fig. 19 (photo of type-specimen).

Material Examined.—GERDA station G-170, Straits of Florida, E of Hobe Sound, Florida (27°06′N, 79°32′W, 677-659 meters, 29 June 1963; one specimen, length 18.1 mm, width 11.6 mm).—G-1082, Straits of Florida, SE of Marquesas Keys (24°24.5′N, 82°02.5′W, 115 meters, 26 April 1969; one specimen, length 30.6 mm, width 18.2 mm).—PILLSBURY Sta. P-605, northwestern Caribbean Sea, NW of Chinchorro Bank, Yucatan (18°50.1′N, 87°31.5′W, 695-772 meters, 17 March 1968; one specimen, length 26.2 mm, width 16.6 mm).

Remarks.—Direct comparison of *Coralliophila lamellosa* (Philippi) from

FIGURE 51. Coralliophilidae. *Coralliophila lamellosa* (Philippi), Sta. P-605, length 26.2 mm.

Sardinia (USNM No. 192734, Tiberi Coll.) with *Coralliophila lintoni* from south of Cuba (USNM No. 93243, ALBATROSS Sta. 2129, 274 fm) and the holotype of *Trophon lintoni* Verrill & Smith from ALBATROSS Sta. 1118 (USNM No. 77269) reveals no significant differences. Moreover, all agree satisfactorily with the present material. The specimen from GERDA Sta. G-1082 is very similar to Verrill & Smith's type of *T. lintoni*. Those from Stas. G-170 and P-605 are somewhat more shouldered, as in Dall's *C. deburghiae* var. *lintoni* from the Caribbean.

43. *Coralliophila lactuca* Dall, 1889
Fig. 52

Coralliophila lactuca Dall, 1889a, Bull. Mus. comp. Zool. Harv., *18*: 220, pl. 16, fig. 6; 1889b, Bull. U. S. Nat. Mus., *37*: 122, pl. 16, fig. 6.

Material Examined.—GERDA Sta. G-67, Straits of Florida, off Cape Florida (25°31'N, 79°57'W, 351 meters, 26 September 1962; one dead shell, length 11.9 mm).—G-179, Straits of Florida, NW of Little Bahama Bank (27°41'N, 79°11'W, 549-567 meters, 1 July 1963; one live specimen, length 10.4 mm).

FIGURE 52. Coralliophilidae. *Coralliophila lactuca* Dall, Sta. G-179, length 10.4 mm.

Description.—See Dall, 1889a: 220.

Remarks.—This distinctive species seems to have been taken rather infrequently. Dall reported it from BLAKE station 5 off the coast of Cuba in 152-229 fathoms (= 274-412 meters), and from ALBATROSS station 2669 off Fernandina, Florida, in 352 fathoms (= 634 meters). It has been taken only twice in the exploratory work of R/V GERDA, both within the range established by Dall.

Family Buccinidae

44. *Phos beauii* Fischer & Bernardi

Fig. 53

Phos Beauii Fischer & Bernardi, 1860, J. Conchyliol., *5*: 358, pl. 12, fig. 8-9 ("Marie-Galante. Trouvé sur les nasses des pêcheurs.").—Tryon, 1881, Man. Conch., *3*: 219, pl. 84, fig. 533.

Phos Beaui, Dall, 1889a, Bull. Mus. comp. Zool. Harv., *18*: 178 (Barbados).

Record.—PILLSBURY Sta. P-875, off St. Vincent (13°10.2′N, 61°5.5′W, 105-183 m, one specimen, length 41.7 mm).

FIGURE 53. Buccinidae. *Phos beauii* Fischer & Bernardi, Sta. P-875, length 41.7 mm.

Family Turbinellidae

Genus *Teramachia* Kuroda

Mesorhytis, Dall, 1889a, Bull. Mus. comp. Zool. Harv., *18*: 172.—Not Meek, 1876, Rep. U. S. Geol. Surv. Territories, *9*: 356, 364 (type-species, *Fasciolaria? gracilenta* Meek).

Teramachia Kuroda, 1931, Venus, *3*(1): 45-47 (type-species, *T. tibiaeformis* Kuroda, by monotypy).—Weaver & du Pont, 1970, Living Volutes: 176.

A living specimen of the peculiar gastropod called *Mesorhytis meekiana* by Dall and another belonging to a distinct but evidently related species were collected in the Caribbean Sea by R/V PILLSBURY. The fact that

196   *Tropical American Mollusks*

FIGURE 54. Turbinellidae. Left, *Teramachia meekiana* Dall, Sta. P-1225, length 26 mm; right, *Teramachia chaunax*, n. sp., Sta. P-904, holotype, length 28 mm.

only two specimens have been obtained over a period of more than six years from among thousands of gastropods trawled in the tropical western Atlantic during the Deep-Sea Biology Program of the Rosenstiel School of Marine and Atmospheric Sciences attests to their considerable rarity.

Dall (1889a) placed *Mesorhytis meekiana* in the family Fasciolariidae, but had neither operculum nor radula for guidance. However, comparison of specimens with *Teramachia barthelowi* (Bartsch) shows such close similarity of conchological characters of these two species that systematic relationship is suggested even in the absence of information about radular and opercular characters of the latter. These features in "*Mesorhytis*" *meekiana* are neither volutid nor fasciolariid, but indicate a closer affinity with the family Turbinellidae. Whether small shells such as *Teramachia barthelowi* eventually will prove to be congeneric with larger forms such as *Teramachia tibiaeformis* Kuroda (the type-species) and *T. mirabilis* (Clench & Aguayo) (the type-species of *Howellia* Clench & Aguayo) must remain for future research to demonstrate. However, the striking similarity of *T. barthelowi* with "*Mesorhytis*" *meekiana*, here placed in the Turbinellidae, is sufficient evidence to assign them both to *Teramachia* along with the new species here described. *Mesorhytis costatus* Dall, 1890, is placed in the same genus on the basis of strong similarity of the shell.

45. *Teramachia meekiana* (Dall)
Figs. 54 (left); 55, D-E

Fasciolaria (*Mesorhytis*) *Meekiana* Dall, 1889a, Bull. Mus. comp. Zool. Harv., *18*: 172, pl. 36, fig. 7; 1889b, Bull. U. S. natn. Mus., *37*: 112, pl. 36, fig. 7 (no descr.; fig. from 1889a).

Material Examined.—PILLSBURY Sta. P-1225, off SW coast of Jamaica (17°42.5'N, 77°58'W, 549-530 meters, 6 July 1970; one specimen).

Description.—This specimen conforms in all essential details with Dall's original material, consisting of dead but fresh specimens from three localities; the figured specimen is 15.5 mm long and 5.0 mm broad. The present specimen is 26.0 mm long and 8.2 mm broad, therefore possibly mature. The apex is lost and there are six remaining postnuclear whorls. Strong axial ribs are present on the second, third, and fourth whorls, becoming obsolete on the fifth. These whorls also show fine, spiral threads most distinct on the upper half, one of which is stronger than the rest. All fade away in the fifth whorl. The body whorl is somewhat more inflated than in Dall's illustrated specimen, the outer lip slightly flared and the anterior canal more distinctly recurved, features probably associated with maturity. The parietal wall has an extremely thin, inconspicuous glaze but is not callused. The columella is moderately flexuous, with three high, narrow, oblique plaits of which the posterior (upper) one is strongest.

The operculum (Fig. 55, E) is translucent corneous yellow, small, thin, and delicate. It is spatulate in shape, weakly curved, with nearly parallel sides. The nucleus is terminal, the apex truncated, perhaps because of wear; the muscle scar, at the opposite end, is almost circular.

The radula (Fig. 55, D) is triserial. The rachidian has three strong cusps on a bent basal plate. The laterals have a single flattened, curved, clawlike cusp arising terminally from an elongated base.

Syntypes.—USNM No. 86970. BLAKE Sta. 100, off Morro Light, Cuba, 400 fm. Two specimens.

Remarks.—The larger of two syntypes of *M. meekiana* Dall (USNM No. 86970), when compared with the type of *Prodallia barthelowi* Bartsch (from ALBATROSS Sta. 5425 in the Sulu Sea off Cagayan I., 495 fm = 905 meters, USNM No. 238444), is smaller and not so strongly sculptured, and its columellar plications are higher and thinner. The axial sculpture of *P. barthelowi* is conspicuous and consists of sharply defined axial ribs separated by narrower deep channels. This sculpture fades out rather abruptly on the penultimate whorl, consequently the body whorl is smooth except for irregular growth marks. The three columellar plications are low but distinct, originating rather deep within the aperture so they are not conspicuous in direct apertural view. Perhaps it is this fact that caused Weaver & du Pont (1970: 177) to state that plications are absent, although they are visible in the photograph reproduced on their plate 75D.

The specimen from Sta. P-1225 is larger than Dall's specimens of *Mesorhytis meekiana* and is approximately the same size as the type of *Prodallia barthelowi*, differing from it in the same features. The conchological characters correspond so closely, however, that there can be little doubt that the two species are congeneric.

46. **Teramachia chaunax**, new species
Figs. 54 (right); 55, B-C

Description.—The shell is elongate fusiform, rather strong, smooth, white suffused with pale brown under a thin, olivaceous periostracum. Length of shell 28 mm, width 8.4 mm. Apex lost, eight postnuclear whorls remaining. Strong axial ribs on the spire, persisting on the body whorl but somewhat weaker there. Twelve axials on the first postnuclear whorl, increasing to 16 on the body whorl. Distinct axial growth lines are present in addition to the axial ribs. A single spiral cord below the suture, most distinct on the early whorls, producing a low nodule where it crosses each axial, thus giving the whorls a faintly shouldered aspect, growing indistinct on the body whorl. Outer lip simple, sharp; parietal wall smooth, not callused. Columella nearly straight, with three strong oblique plaits, the uppermost one strongest, forming deep within the aperture and scarcely visible from without. Anterior canal slightly produced, with about a dozen weak spiral threads.

The operculum (Fig. 55, C) is elongate, rather strongly curved, with terminal nucleus; the muscle scar is ovate and rather large.

The radula (Fig. 55, B) is triserial, as in *T. meekiana*, but the rachidian has a large, erect cusp with usually three denticles on each side. The laterals have a flattened, clawlike cusp at the end of an elongated base, as in *T. meekiana*.

Holotype.—USNM No. 701216, PILLSBURY Sta. P-904.

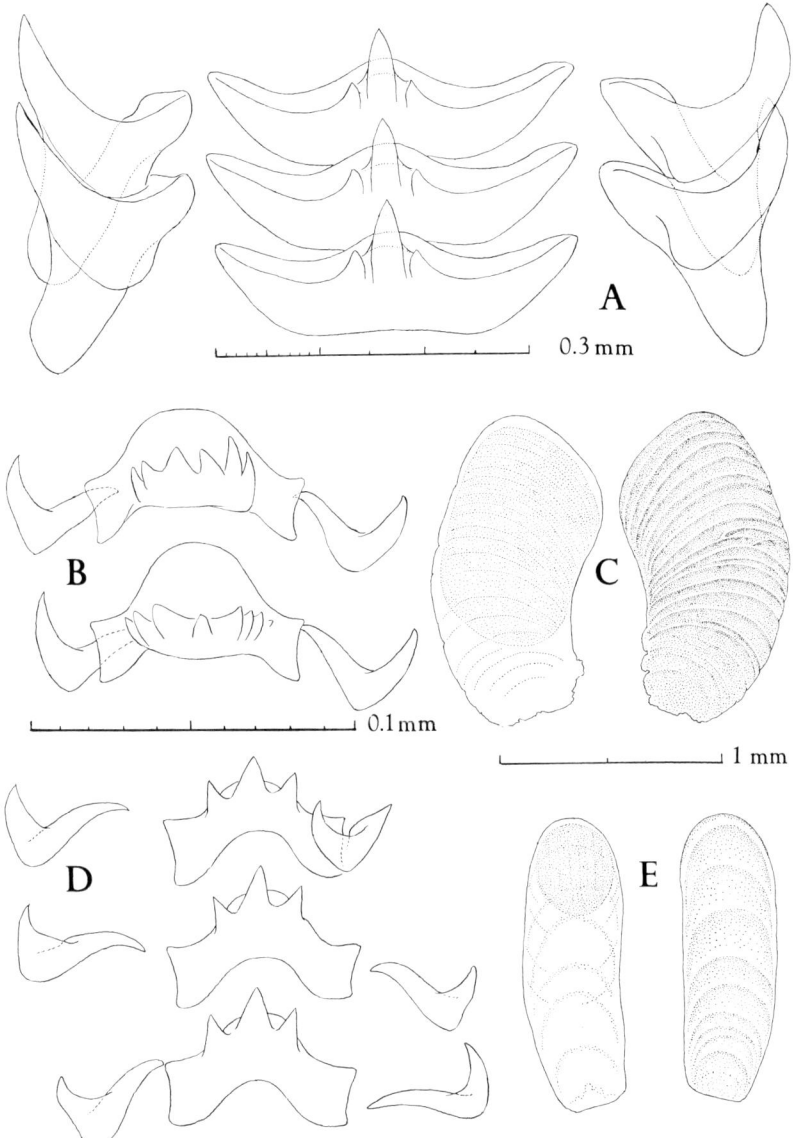

FIGURE 55. Turbinellidae. A, *Turbinella laevigata* Anton, OREGON Sta. 4240, radula; B, *Teramachia chaunax*, n. sp., Sta. P-904, radula; C, operculum of same; D, *Teramachia meekiana* Dall, Sta. P-1225, radula; E, operculum of same.

Type-Locality.—PILLSBURY Sta. P-904, W of St. Lucia (13°45.5'N, 61°05.7'W, 201-589 meters, 9 July 1969).

Remarks.—This shell is very much like Dall's *M. meekiana* as originally described and figured, but differs in the persistence of the axial ribs on all whorls, the reduction of the subsutural spirals to a single cord at the shoulder, and the origin of the columellar plaits very deep in the aperture. The size of the aperture relative to the height is more like that of the original material of *M. meekiana*, which apparently was not adult, than that of the larger specimen from Sta. P-1225.

47. *Turbinella laevigata* Anton
Fig. 55, A

Turbinella laevigata Anton, 1839, Verzeichniss der Conchylien: 71.
Xancus laevigatus, Abbott, 1950, Johnsonia, 2(28): 207, pl. 91.

Record.—OREGON Sta. 4240, off coast of Maranhão State, Brazil, north of Tutoía: 2°04'S, 42°05'W, 49 meters, bottom temp. 83°F, 11 March 1963. Nine specimens.

Remarks.—The radula of this species is illustrated for comparison with those of *Vasum* and "*Mesorhytis*" (= *Teramachia*). As the species is so infrequently collected, the above locality record is reported.

48. *Vasum capitellum* (Linnaeus)
Fig. 56

Murex Capitellum Linnaeus, 1758, Syst. Nat., Ed. 10: 750.
Vasum (Altivasum) capitellum, Abbott, 1950, Johnsonia, 2(28): 214, pl. 94.
Vasum capitellus, Warmke & Abbott, 1961, Caribbean, Seashells: 121, pl. 22q.—de Jong & Kristensen, 1965, Correspondentieblad Ned. Malacol. Vereniging, Suppl. 1965: 39.
Vasum capitellum, Vokes, 1966, Tulane Stud. Geol., 5(1): 20 (synonymy).
—Work, 1969, Bull. Mar. Sci., 19(3): 675.

Material Examined.—PILLSBURY Sta. P-916, Guadeloupe (16°12.2'N, 61°26.3'W, 2 m, 11 July 1969, two specimens).

Remarks.—This species is not common in collections and was taken only once by R/V PILLSBURY, at a shore station. As the operculum has not been figured and nothing seems to be known of the soft parts, the operculum and gross anatomy of the mantle cavity are illustrated in Figure 56.

Family Volutidae

Collections obtained by R/V GERDA and R/V PILLSBURY include members of the Volutinae, Lyriinae, Scaphellinae, and Volutomitrinae, and provide further evidence bearing upon classification in the family. The western Atlantic species of this family have been described and illustrated admirably

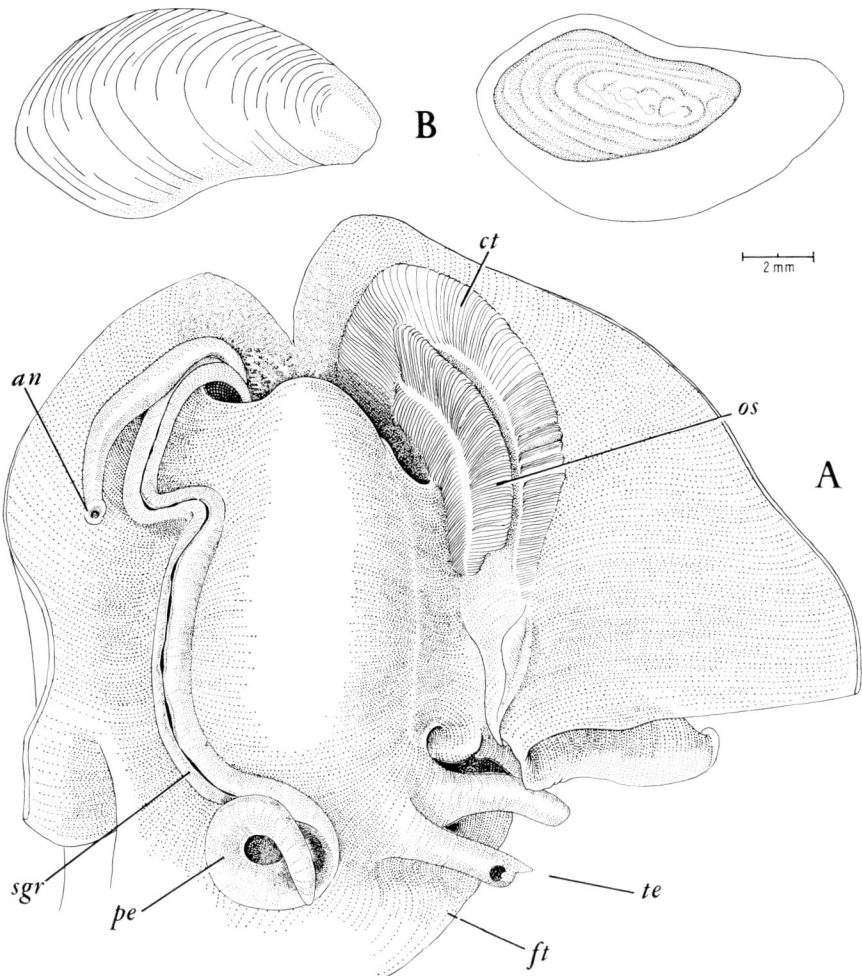

FIGURE 56. Turbinellidae. *Vasum capitellum*: A, gross anatomy of the mantle cavity of a specimen from PILLSBURY Sta. P-916; B, operculum. Drawing of operculum by Constance S. McSweeny. (*an*, anus; *ct*, ctenidium; *ft*, foot; *os*, osphradium; *pe*, penis; *sgr*, seminal groove; *te*, tentacle with eye.)

by Clench (1946, 1953) and Clench & Turner (1964, 1970). The classification of the family, still a difficult matter, has been considered by Pilsbry & Olsson (1953, 1954), who reviewed its history (1954).

The form of the radular teeth has been regarded as a fundamental character in subdividing the family Volutidae. It is hardly open to question that this is a sound concept, so long as the functional modifications that may occur in this structure are kept in mind.

It is difficult to relate, at first glance, the comblike rachidian plate of *Voluta* with the strongly modified tricuspid teeth of *Volutifusus, Aurinia* and *Odontocymbiola*, and the unicuspid teeth of *Scaphella*. However, intermediate stages can be seen even in the limited material examined in the present study, and these may be related to the development of specialized feeding habits.

The multicuspid rachidian is comblike in *Voluta musica* (Clench & Turner, 1964) but in *V. virescens* it shows a definite tendency toward a tricuspid condition by the clear emphasis of the middle as well as the two lateral cusps (Olsson, 1965, pl. 83, fig. 6a; Clench & Turner, 1970, pl. 173, fig. 4; and the present work, Fig. 60, B). This trend is continued in teeth such as those of *Lyria cordis*, in which the cusps between the central and lateral ones are reduced to denticles (Fig. 60, A). Loss of these denticles and increasing curvature of the basal plate would result in teeth such as those of *Zidona* (Clench & Turner, 1964, pl. 91) and *Aurinia* (Fig. 63, A). Reduction of the lateral denticles would result in teeth of the *Clenchina* and *Scaphella* types (Figs. 63, C; 63, B).

The extremely small size of the radulae in the Scaphellinae would apparently make the structure ineffective as a rasping organ, but the distinctly channeled cusp quite conceivably could serve to introduce venomous secretions into a prey animal. *Scaphella junonia* observed in the aquarium very quickly subdue their prey before transferring it to the pouch-like posterior end of the foot, in which it is kept (alive?) until it is eaten. The prominent cusp of *S. junonia* is conspicuously gutterlike.

The strongest suggestion that these kinds of teeth are assuming a fanglike function can be seen in the exceedingly small teeth of *Volutomitra*, in which the cusp is dartlike, sharply pointed, deeply channeled, and has very sharp edges. In these teeth, the basal shanks are strongly developed, apparently to provide greatest support for the dartlike cusp. This tooth form evidently has developed in somewhat the same way as has that of *Odontocymbiola* (see Clench & Turner, 1964, pl. 109; also Pilsbry & Olsson, 1954, pl. 4, figs. 6, 6a, 6b); published figures suggest that at least the central cusp may be channeled or hollow.

It is evident that studies of the feeding behavior of volutids are sorely needed as an aid toward understanding the evolution of radular form.

49. *Voluta virescens* Lightfoot
Figs. 57; 60, B

Voluta virescens Lightfoot, 1786, Cat. Portland Mus., London: 26, No. 610.
—Olsson, 1965, Bull. Amer. Paleont., *49*(224): 661, pl. 80, figs. 6-6a

FIGURE 57. Volutidae. *Voluta virescens* Lightfoot, Sta. P-324, length 54.1 mm.

(type of *V. fulva* Lam.), 7-7a (*V. polyzonalis* Lam. from Lamarck's collection); pl. 81, figs. 1-1a (type of *V. polyzonalis* Lam.), 2-2a (specimen figured by Reeve, Conch. Icon.), 6 (fossil specimen); pl. 82, fig. 5 (spire of Recent specimen), 6 (spire of fossil specimen); pl. 83, figs. 1-1a, 2 (Recent specimens, Panama), 3 (fossil, Panama), 6-6a (radula).—Clench & Turner, 1970, Johnsonia, 4(48): 370, pl. 173, fig. 4 (radula).—Weaver & du Pont, 1970, Living Volutes: 7, pl. 1, figs. A, B.
Voluta? *virescens*, Clench & Turner, 1964, Johnsonia, 4(43): 146, pls. 82, 84.

This species, generally considered rare, was taken by R/V PILLSBURY at ten stations along the Caribbean coast of Panama and Colombia.

Records.—PILLSBURY stations P-324, Panama (9°44'N, 79°31'W, 64-65 m; 1 live specimen 54 mm long, 26 mm wide).—P-348, mouth of Gulf of Urabá (8°38.0'N, 77°02.2'W, 59 m; 1 live specimen 89.2 mm long, 42.7 mm wide).—P-367, Colombia (9°31.1'N, 75°49.6'W, 37-35 m; 2 small broken specimens, possibly preyed upon by some crustacean).—P-368, Colombia (9°31.2'N, 75°41.1'W, 37 m; six live specimens from 40.2 mm long, 18.2 mm wide, to 24.8 mm long, 12.8 mm wide).—P-370, Colombia

(9°37.9'N, 75°50.4'W, 37 m; one dead specimen 46.5 mm long, 23.5 mm wide; and 1 live specimen 32.3 mm long, 16.2 mm wide).—P-396, Colombia (9°18.2'N, 76°24.8'W, 70-68 m; one dead specimen 55 mm long, 25.8 mm wide).—P-397, Colombia (9°12.8'N, 76°27.1'W, 62-66 m; one dead specimen 28.7 mm long, 15.9 mm wide).—P-403, mouth of Gulf of Urabá (8°48.7'N, 77°12.7'W, 99-97 m, one live specimen 15 mm long, 8.4 mm wide).—P-412, Panama (8°38.9'N, 77°13.2'W, 55-60 m; one dead specimen 31.5 mm long, 15.4 mm wide).—P-434, Golfo de los Mosquitos, Panama (9°14.6'N, 80°21.8'W, 49-48 m; portion of body whorl of adult specimen).

Remarks.—The largest specimen (P-348, 89.2 mm long) is similar in size and shape to that illustrated by Reeve (Olsson, 1965, pl. 81, figs. 2-2a), which shows a tendency toward the growth form called *polyzonalis* Lam.; unfortunately, its shell is badly disfigured by scars from old shell damage and by the heavy blackish deposits commonly seen on mollusks in the southern Caribbean. The smallest example is a juvenile only 15 mm long, with four whorls not clearly divisible into nuclear and postnuclear turns; axial ridges begin weakly after a little more than two full turns and are strong on the fourth (last) whorl. The lot of six specimens from P-368, all young, contains both the strongly shouldered, low-spired form and the narrower, higher-spired form; color and sculpture are identical.

The radula of a young animal from P-368 (Fig. 60, B) agrees more closely with the figure given by Olsson (1965) than with that of Clench & Turner (1970). It is composed of 50 teeth (the last two incompletely formed) 0.24 mm wide. The central and terminal cusps are stronger than the others, showing a trend toward the tricuspid condition of the Scaphellinae. A possible intermediate stage is seen in the teeth of *Lyria cordis*, n. sp., in which the cusps separating the central and terminal ones are reduced to small denticles (Fig. 60, A).

Genus *Lyria* Gray

Subgenus **Cordilyria**, new subgenus

Diagnosis.—Shell smooth, scaphelliform, axial sculpture developed on early whorls but becoming obsolete toward the body whorl; columella with three strong plaits followed by conspicuous lirations on the parietal wall. Radula conspicuously tricuspid, interspaces between cusps denticulate. Operculum well developed, nucleus apical.

Type-Species.—*Lyria* (*Cordilyria*) *cordis*, new species, here designated.

50. **Lyria (Cordilyria) cordis**, new species

Figs. 58; 59; 60, A; 61

Description.—Shell ovately fusiform, smooth and glossy but not highly polished. Anterior canal weakly produced, the small sinus forming a fas-

FIGURE 58. Volutidae. *Lyria cordis*, n. sp., holotype, Sta. P-1303, length 42.5 mm.

ciole indicated chiefly by direction of growth lines and slightly paler color, not distinctly bounded by any sculptural cord or ridge. Length 42.5 mm; maximum diameter 18 mm; length of body whorl 27 mm. Whorls seven, of which 1½ compose the small, smooth protoconch, which is not sharply distinguished from the postnuclear whorls; the following 2½ whorls sculptured by low, rounded axial undulations which slowly diminish on the next two whorls, until the body whorl becomes quite smooth, marked only by microscopic axial growth lines. Spiral sculpture virtually nonexistent, the postnuclear whorls showing only the faintest indication of widely spaced spiral striation. Outer lip slightly thickened, not denticulate within. Columella with three prominent folds, the middle one strongest, the posterior one weakest; parietal wall with 12 narrow, raised ridges, the anterior ones much weaker than the adjacent columellar fold, the posterior ones very faint.

Ground color vinaceous tan, suture with a diffuse pale spiral line, faint indication of a pale spiral band below the suture, another below the middle of the body whorl, and one along the fasciole. A color pattern of reddish brown spots begins about the third whorl, becoming more distinct on the succeeding whorls; body whorl with eight spiral rows of bilobed, heart-shaped, or squarish spots arranged in axial rows not strictly conforming with the contour of the lip. Four marginal spots clearly visible through the edge of the lip and a fifth (the anteriormost) less distinctly so. A small area of whitish callus lies adjacent to the posterior angle of the aperture; parietal area without color pattern, as if the outermost layer of shell has been planed off by the advancing edge of the mantle. This color pattern results in a shell resembling a small *Scaphella* or a spotted *Aurinia*.

The gross external features of the soft parts agree with the description of those of *Voluta* given by Clench & Turner (1964). The siphon has a pair of unequal basal lobes, the left one the larger. The head is strongly flattened, with a broad frontal lobe lying between the tentacles. The tentacles are flattened, tapering, the right one broader than the left. A broad lateral lobe lies on each side of the head behind the tentacles, the right one larger than the left. The eyes are situated near the anterior margin of the lateral cephalic lobes near their junction with the tentacles. The broad, flattened proboscis extends from beneath the frontal lobe. The operculum is situated on a prominent opercular pad on the dorsal side of the foot near its posterior end.

The color of the living animal is shown in Figure 59.

The uniserial radula has a tricuspid rachidian with a series of small, but distinct, denticulations between the cusps (Fig. 60, A).

Holotype.—USNM No. 700000, from PILLSBURY Sta. P-1303.

Paratype.—USNM No. 700001, from the same station.

FIGURE 59. Living animals of *Lyria cordis*, n. sp. (above) and *Ficus howelli* Clench & Aguayo (below). Photographs by Dennis M. Opresko.

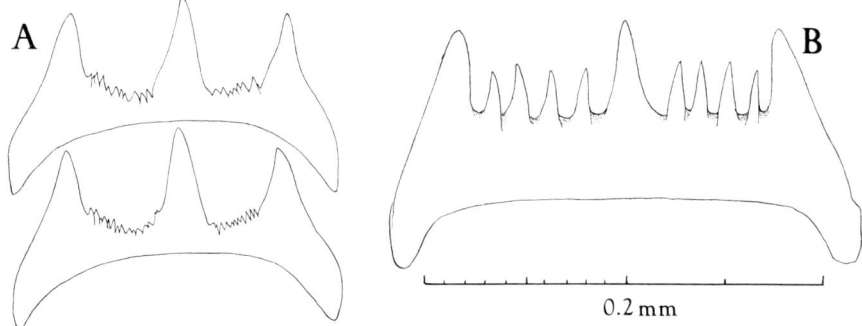

FIGURE 60. Radulae: A, *Lyria cordis*, n. sp.; B, *Voluta virescens* Lightfoot.

Type-Locality.—PILLSBURY Sta. P-1303, Caribbean Sea, 20 miles ESE of Sto. Domingo, island of Hispaniola, 18°21.0'N, 69°14.3'W, depth 174 m, 21 July 1970.

Remarks.—By its smooth shell, *Lyria cordis* resembles *Lyria* (*s.s.*) *vegai* Clench & Turner, 1967, also from Hispaniola, but its coloration and shape differ markedly. The aperture is longer and narrower, occupying 64 per cent of the total length in contrast to 56 per cent in *L. vegai* (calculated from published photograph), resulting in a more "scaphelloid" shape. This *Scaphella*-like appearance is accentuated by the color pattern of reddish brown spots arranged in regular spiral rows. The columella has three well-developed plications and a dozen parietal ridges, compared with two strong and one weak plication and only three or four parietal ridges in *L. vegai*. *L. cordis* has a smaller shell, evidently maturing at a little over 40 mm rather than reaching 60 mm as in *L. vegai*. In its general shape, *Lyria cordis* seems to approach *L. beaui* (as figured by Fischer & Bernardi, reproduced by Clench & Turner, 1967: figs. 2-3) more closely than does *L. vegai*, but Fischer & Bernardi's figure seems to be a little misleading as shown by modern colored photographs (Dance, 1969: 96, pl. 16c). *Lyria cordis* lacks axial costae on the later whorls.

The radula of *L. vegai* is unknown, but those of *L. beaui* (Tryon, 1882: 101, pl. 2, fig. 7) and of the Indo-Pacific *L. mitraeformis* Lamarck and *L. quecketti* Smith (Cooke, 1922: 9, figs. 5, 6) lack denticles between the cusps, a feature that sets *L. cordis* apart from all other species of *Lyria* whose radulae have been described.

The discovery within a relatively short time of two very showy new species from the same geographical region is another indication of our very incomplete knowledge of the Caribbean molluscan fauna.

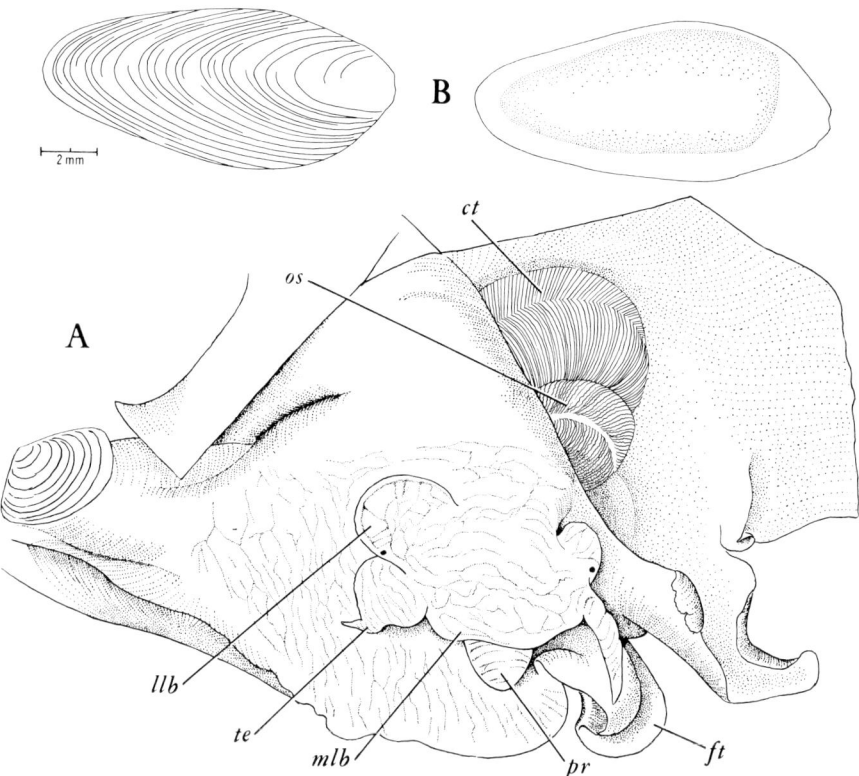

FIGURE 61. Volutidae. *Lyria cordis*, n. sp.: A, gross anatomy of the mantle cavity; B, operculum. Drawing of operculum by Constance S. McSweeny. (*ct*, ctenidium; *ft*, foot; *llb*, lateral lobe of head; *mlb*, median lobe of head; *os*, osphradium; *pr*, proboscis; *te*, tentacle.)

Genus *Scaphella* Swainson, 1832

Pilsbry & Olsson (1953) realigned the species of the subfamily Scaphellinae on the basis of radular characters, clearly distinguishing three types of rachidian teeth: (1) those with a single long, concave cusp and no small basal denticles (genus *Scaphella*); (2) those with a shorter, more pointed, concave cusp flanked by minute accessory cusps (genus *Clenchina*); and (3) those with a well-developed lateral cusp on each side of the main, central cusp (genera *Aurinia* and *Volutifusus* = *Bathyaurinia*). Those authors observed that the accessory cusps in *Clenchina* are the sharply pointed ends of the two ridges along the sides of the basal shanks of the tooth.

Careful examination of the teeth of *Scaphella junonia* shows that the ends of the lateral ridges of the shanks are produced as inconspicuous rounded "shoulders" at the base of the cusp, so the accessory denticles are present, although in a reduced condition, also in *Scaphella junonia*. It appears, therefore, that the teeth in this subfamily show a progressive reduction of the lateral denticles from the strongly tricuspid condition of *Aurinia* to the simple Y-shape in *Scaphella*, where the lateral denticles have nearly disappeared. This fact, together with the similarity of conchological characters, seems to justify recognition of the groups established by Pilsbry & Olsson at the subgeneric rather than the generic level. The differences are small, and information about the radulae of more species may well erase the distinctions altogether.

51. *Scaphella* (*Scaphella*) *junonia* (Lamarck)
Figs. 62; 63, B

Voluta junonia Lamarck, 1804, Ann. Mus. natn. d'Hist. Nat., 5: 156.

Scaphella (*Scaphella*) *junonia*, Clench, 1946, Johnsonia, 2(22): 49, pl. 28, figs. 1-3 (synonymy).—Clench & Turner, 1970, Johnsonia, 4(48): 371 (revised distribution).—Weaver & du Pont, 1970, Living Volutes: 140, figs. 31a, 31b; pl. 57 E-H.

Record.—GERDA Sta. G-767. Straits of Florida, SSE of Carysfort Reef Lighthouse: 25°13'N, 80°10'W, 108-88 meters, 26 January 1966, one specimen, length 98.5 mm, width 39 mm at time of capture; length 107.4 mm, width 46.5 mm on June 5, 1967, after 16 months in the aquarium.

Range.—North Carolina southward to the Straits of Florida; Gulf of Mexico, including the Bay of Campeche, to Arrowsmith Bank, Yucatan.

Remarks.—The radular teeth of the specimen from Sta. G-767 (Fig. 63, B) agree with the figures given by Pilsbry & Olsson (1953: figs. 1, 1a; 1954: pl. 3, figs. 14, 14a), but not with that of Clench (1946: pl. 24, fig. 4), which shows the basal shanks extending back almost parallel from the long, straight cusp.

This specimen was maintained alive in running seawater at the institute for a period of 16 months by Mr. Robert C. Work. During this period, the animal added 29 mm to the margin of the shell, resulting in an increase in length of almost 10 mm and in maximal width of over 7 mm. The spots on the new part of the shell are more intensely colored, and those in four of the spiral rows subdivided into two rows each (Fig. 62). This shows that environmental factors can influence the number of rows, size, and color of the spots and suggests that nominal subspecies such as *S. junonia butleri* Clench and *S. junonia johnstonae* Clench are localized populations whose color characters are influenced by ecological conditions. The fol-

FIGURE 62. Volutidae. *Scaphella junonia* (Lamarck), Sta. G-767, length 107.4 mm.

lowing observations on the activities of this animal were generously provided by Mr. Work:

> Very soon after its release on a water-table, the junonia pursued and captured a naticid that had been taken in the same dredge haul. The prey was completely and tightly enfolded by the posterior portion of the foot, which appeared to form a deep pouch opening to the outside only through a single ventral, anterior slit. The junonia then curled its anterior portion ventrally and posteriorly, bringing its head very near the entrance to the pseudopouch formed by its foot. As the victim was thought to be an undescribed species, it was rescued immediately and showed no ill effects from the attack.
>
> During the subsequent 28 days, the junonia made no further attempt to capture prey, even though specimens of many different genera of gastropods and pelecypods, as well as other invertebrates, were freely available. On the 29th day of its captivity, living specimens of *Oliva sayana* and *Cancellaria reticulata* were released on the water table with the junonia. It promptly attacked a large olive, repeating the method of enfoldment that had been employed with the naticid. The olive shell was enfolded with its long axis at right angles to the longitudinal axis of the junonia's foot, so that the pouch protruded laterally beyond the usual width of the foot.

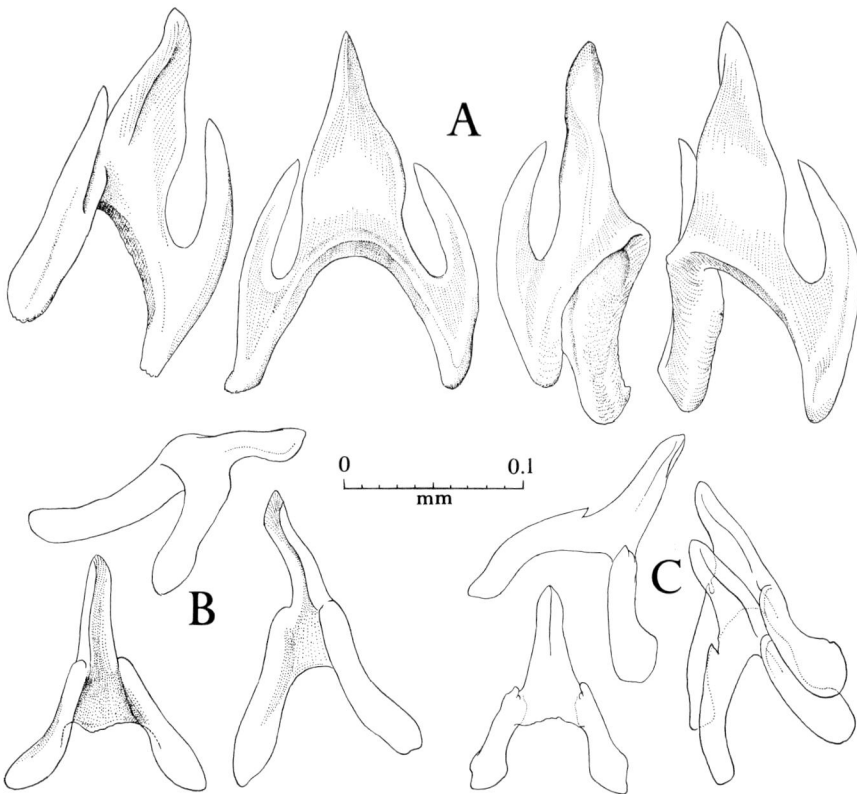

FIGURE 63. Radular teeth: A, *Scaphella (Aurinia) dubia*; B, *Scaphella (Scaphella) junonia*; C, *Scaphella (Clenchina) evelina*, n. sp.

In subsequent attacks on *Oliva*, the prey was always held in the same manner. The junonia then lay on its side and inserted its head into the pouch for a short time, after which the olive no longer showed any indication of struggle. The junonia then dragged its victim around the water-table for several hours, making no attempt to eat it. The olive was then retrieved, and appeared to be dead but completely intact. It was then replaced near the junonia, which immediately enfolded it and resumed dragging it. Subsequent retrieval of the olive showed that a large portion of the soft parts had been eaten. The same olive was again accepted by the junonia and dragged around the water-table for a period of at least eight hours and possibly much longer into the night. The next morning, the junonia was found attacking a specimen of *Cancellaria*. This animal also appeared to be stunned or dead and was dragged about for a long period before it was eaten. For several more months the junonia was kept on the water-table, where it regularly attacked and dragged olives about, intermittently stopping to feed on its prey "in storage."

As the shallow sand of the water table was insufficient for the junonia to burrow deeply, it is possible that this was the reason it dragged its prey for long periods. When the animal later was placed in a large tank with deeper sand, where it remained until it was preserved more than a year later, it captured and devoured all of its prey beneath the surface of the sand, making accurate observation impossible. A constant supply of *Oliva* was maintained in the tank. During the entire period of captivity, only *Oliva*, *Cancellaria* and *Natica* were attacked by the junonia.

52. Scaphella (Aurinia) dubia (Broderip, 1827)
Fig. 63, A

Voluta dubia Broderip, 1827, Zool. Jour., *3*: 81, pl. 3, fig. 1.
Scaphella (Aurinia) dubia, Clench, 1946, Johnsonia, 2(22): 54, pl. 30, figs. 1-2 (synonymy).—Weaver & du Pont, 1970, Living Volutes: 144, pl. 29, figs. C, D.

Record.—GERDA Sta. G-462. Straits of Florida, SE of Dry Tortugas (24°20'N, 82°46'W, 174-201 meters, 25 January 1965, two specimens, length 68 mm, width 23.4 mm; length 70.4 mm, width 21.7 mm).

Range.—Gulf of Mexico and lower Straits of Florida.

Remarks.—This record of the type-species of *Aurinia* is introduced for comparison of the radular teeth with those of *Scaphella* (*Clenchina*) *evelina*, n. sp., and of *Scaphella* (*s.s.*) *junonia* (Lamarck).

53. Scaphella (Clenchina) evelina, new species
Figs. 63, C; 64

Material Examined.—PILLSBURY Sta. P-394, Caribbean Sea, off Isla Fuerte, Colombia (9°28.6'N, 76°26.3'W, 421-641 meters, 16 July 1966; one live specimen, length 91.6 mm, width 34.8 mm).—P-399, Caribbean Sea, NE of Punta Caribana, Colombia (9°01.3'N, 76°40.2'W, 119-179 meters, 17 July 1966; large fragment of columella and body whorl of large specimen).—P-444, Caribbean Sea, Golfo de los Mosquitos, Panama (8°57.5'N, 81°31.0' W, 73 m, 21 July 1966; one dead specimen, length 110.9 mm, width 42.5 mm).—P-797, Caribbean Sea, off Cartagena, Colombia (10°21.9'N, 75°47.3'W, 170-150 m, 1 August 1968; one dead specimen, juvenile, length 23.8 mm, width 11.2 mm).—OREGON Sta. 3587, Caribbean coast of Panama (9°18'N, 80°25'W, 137 m, 29 May 1962; one dead and broken specimen, length 181 mm, width 76 mm).

Description.—Shell large, fusiform, first three postnuclear whorls thick and solid, but shell becoming thinner with increasing size; body whorl of mature shells thin and delicate, rather inflated when fully grown. Whorls six,

FIGURE 64. Volutidae. *Scaphella evelina*, n. sp.: left, holotype, Sta. P-394, length 91.6 mm; right, paratype, Sta. P-445, length 110.9 mm.

weakly shouldered from the third postnuclear turn onward, with a faintly concave slope above the rounded shoulder. Spire acute, moderately extended, suture moderately impressed. Aperture elongate-elliptical, narrowed above and more so toward the siphonal canal. Outer lip thin and fragile, parietal wall thinly glazed. Columella weakly arched, with two weak plications visible deep within the aperture (juvenile of only two postnuclear whorls with three plications, suggesting that the anterior one progressively weakens and ultimately forms the columellar edge of the anterior canal). Siphonal canal broad, gently tapered, only little arched dorsally. Nuclear whorl smooth; first postnuclear whorl developing spiral cords which become strong in the second whorl, where weakly curved axial cords appear, producing a distinct beading where they intersect with the spirals; axial riblets becoming about as strong as the spiral cords in the third postnuclear whorl, so the sculpture appears cancellate; after third whorl, sculpture weakening, the axials regressing to microscopic growth lines, the spirals scarcely perceptible; surface of body whorl macroscopically smooth. Color of nuclear whorl brown; ground color of postnuclear whorls pinkish brown or mahogany brown, with faint, irregular axial streaks; one example with about 12 spiral rows of spirally elongated mahogany brown spots showing a tendency to fuse axially into irregular dark streaks; juvenile of two postnuclear whorls with eight spiral rows of distinct squarish brown spots.

Holotype.—USNM No. 701217, from PILLSBURY Sta. P-394.

Paratypes.—All remaining material; from PILLSBURY Stas. P-399, P-444, P-797, and OREGON Sta. 3587.

Type-Locality.—Caribbean Sea off Isla Fuerte, Colombia, at PILLSBURY Sta. P-394 (9°28.6'N, 76°26.3'W, in 421-641 meters).

Range.—Southwestern part of the Caribbean Sea, at depths between 77 and 641 meters.

Remarks.—This species is distinguished by the cancellate sculpture, produced by the intersection of spiral and axial cords on the first three postnuclear whorls, which fades out to a smooth surface on the body whorl; by its large size, thin structure and, when fully developed, inflated body whorl; and by its coloration. It seems very probable that the uniform brown phase results from fusion of the dark blotches. The close similarity of shell characters in all specimens indicates that only a single species is involved here.

The cancellate sculpture on the spire resembles the condition in *Scaphella junonia* (Lamarck) and *S. floridana* (Heilprin), in which, however, it does not persist to so late a stage in the development of the postembryonic shell (Pilsbry & Olsson, 1954, pl. 1, figs. 5, 9). This conchological relationship is further reinforced by the close similarity of the radular teeth of *S. evelina*, n. sp. (Fig. 63, C) with those of *S. junonia* as illustrated by Pilsbry & Olsson (1953) and Pilsbry & Olsson (1954), and confirmed by material collected by R/V GERDA (Fig. 63, B). The most conspicuous difference is the presence of a small, sharp denticle on each side of the base of the large central cusp, in which it agrees with species of *Clenchina* Pilsbry & Olsson. As indicated above, this genus is here treated as a subgenus of *Scaphella*.

Genus *Volutomitra* (Gray Ms.) H. & A. Adams

Volutomitra Gray, *in* H. & A. Adams, 1853, Gen. Rec. Moll., *1*: 172.—
Gray, 1857, Guide: 36.—Tryon, 1882, Man. Conch., *4*: 108, 124.—
Dall, 1889a, Bull. Mus. comp. Zool. Harv., *18*: 145.

Diagnosis.—Shells of moderate size, fusiform, smooth or weakly sculptured, with thin, smooth periostracum; aperture narrow, without siphonal notch and fasciole, outer lip simple, sometimes with a slight thickening upon full development. Radula extremely small, with a scaphelloid rachidian having flattened shanks and a sharp, guttered cusp; laterals minute, sliver-like (sometimes absent entirely?). No operculum.

Type-Species.—*Volutomitra groenlandica* (Beck) by subsequent designation: Tryon, 1882: 124, pl. 2, fig. 8 (radula) and pl. 36, fig. 83 (shell); also Dall, 1889a: 145, pl. 34, figs. 6-7.

Remarks.—The subfamily Volutomitrinae was omitted from the family Volutidae by Weaver & du Pont (1970). However, the form of the rachidian tooth, similar to that of *Amoria* (see Weaver & du Pont, 1970: 147-167), speaks strongly for its inclusion in the family, perhaps even in the subfamily Scaphellinae. Although the vestigial lateral teeth are so small as to be easily overlooked, and may indeed be lacking in some cases, the triserial radula is sufficient grounds for retention of the subfamily Volutomitrinae, at least provisionally.

It should be restated that the figure of the radula of *V. groenlandica* showing broad, platelike laterals, reproduced by Tryon (1882) from Troschel, undoubtedly represents the detachment of the shanks from the cusps by excessive flattening under the coverslip (Cooke, 1922: 10).

54. **Volutomitra persephone**, new species
Figs. 65; 67, A-C

Material Examined.—PILLSBURY Sta. P-447, Golfo de los Mosquitos, Panama (9°07.4'N, 81°07.4'W, 664-681 m, 21 July 1966; four specimens). —P-741, off Los Roques, Venezuela (11°47.8'N, 66°06.8'W, 1052-1067 m, 23 July 1968; one dead shell).—P-754, off Pta. Zamuro, Venezuela (11°36.9'N, 68°42'W, 684-1574 m, 26 July 1968; five specimens).

FIGURE 65. Volutidae. *Volutomitra persephone*, n. sp., holotype, Sta. P-447, length 40.7 mm.

Description.—Shell of moderate size, porcellaneous white under a thin, smooth, olivaceous periostracum, fusiform, with spire elevated; siphonal canal scarcely produced and slightly flared anteriorly. Whorls ten, spire flatsided, suture little impressed. Aperture narrowly elliptical. Outer lip thin and smooth, its edge faintly reflected. Parietal wall with a narrow glazed area; columella only slightly arched, with four plications of which the anteriormost is weakest and not visible from outside. The other three plications, visible in the aperture, moderately strong; the middle one strongest, the posterior (uppermost) one next, and the anterior one (i.e., the third from above) least so. No siphonal fasciole. Nucleus apparently of about one whorl, but the eroded surface does not reveal the boundary with postnuclear whorls. Areas having intact surface show narrow, semilunate axial ribs beginning in the second whorl and persisting into the fifth where they become obsolete and disappear. Axial growth lines microscopically fine. No evident spiral sculpture on the spire, but there is a subtle hint of spiral lines possibly confined to the periostracum; these continue on the body whorl where, toward the base, they strengthen to form about 15 slightly raised, flattened, spiral cords that become evident opposite the second (i.e., the strongest) columellar plication.

No operculum. Radula small, with rachidian of generally scaphelloid nature, having a cusp (the "mesocone") resembling a pointed, V-shaped trowel blade (Fig. 67, A). No laterals clearly detectable, but the edge of the ribbon has oblique refractive lines that could represent scars from which laterals have been detached or extremely delicate laterals, *per se*. These are precisely as drawn by William Stimpson for *Volutomitra groelandica* in the figure reproduced by Dall (1889a: pl. 34, fig. 7).

Measurements.—Length 40.7 mm, width 16.1 mm (holotype); length 39.3 mm, width 16.1 mm (paratype, broken shell); length 38.7 mm, width 15.1 mm (paratype, live shell); length 37 mm, width 14 mm (paratype, broken shell); and six smaller paratypes from 11.1 to 19 mm in length.

Holotype.—USNM No. 701218, from PILLSBURY Sta. P-447.

Type-Locality.—P-447, Golfo de los Mosquitos, Panama (9°07.4'N, 81°07.4'W, in 664-681 meters).

Remarks.—The spire of the type-specimen bears a hemispherical leathery egg capsule, possibly belonging to this species, and evidence of the former attachment of two others.

The small specimens from stations P-741 and P-754, ranging from 11 to 19 mm in length, are not separable from the larger individuals on any radular or conchological characters, and therefore are retained within this species.

These shells have a striking resemblance to *Volutomitra alaskana* Dall (1921: 87, pl. 11, fig. 3) in size, shape, and general appearance. The chief differences are the more produced anterior canal and the weaker anterior columellar plication in *V. persephone*.

55. **Volutomitra erebus**, new species
Figs. 66; 67, D-J

Material Examined.—PILLSBURY Sta. P-478, SW of Grenada (11°34.4'N, 62°10.7'W, 598-597 m, 2 August 1966; one specimen).—P-776, N of Guajira Peninsula, Colombia (12°13.3'N, 72°50'W, 408-576 m, 29 July 1968, two specimens).—P-781, N coast of Colombia (11°30.1'N, 73°26.5'W, 567-531 m, 30 July 1968, four specimens).

Description.—Shell of moderate size, porcellaneous white under a thin, smooth, pale olivaceous periostracum, with elevated spire; siphonal canal

FIGURE 66. Volutidae. *Volutomitra erebus*, n. sp., holotype, Sta. P-776, length 35.6 mm.

slightly produced, weakly flared anteriorly. Whorls nine, sides of spire somewhat convex, whorls slightly inflated, suture impressed. Aperture narrowly elliptical. Outer lip simple, smooth, somewhat thickened and a little reflected in the adult. Parietal wall with a moderately wide glazed area; columella almost straight, with three distinct plications, of which the lowest is slightly weaker, followed by a weak low ridge not visible from outside and only faintly discernible deep within the aperture. Nucleus apparently of one whorl, but apex eroded so that its limits, and the place of initiation of sculpture, cannot be seen. Axial ribs clearly present in the third and subsequent whorls, persisting onto the penultimate whorl and becoming obsolete but detectable on the body whorl. Axial growth-lines microscopic. Weak spiral lines microscopically visible on spire; lower half of body whorl with low, but distinct, raised spiral cords beginning opposite the anal end of the aperture; irregular axial wrinkles on the body whorl

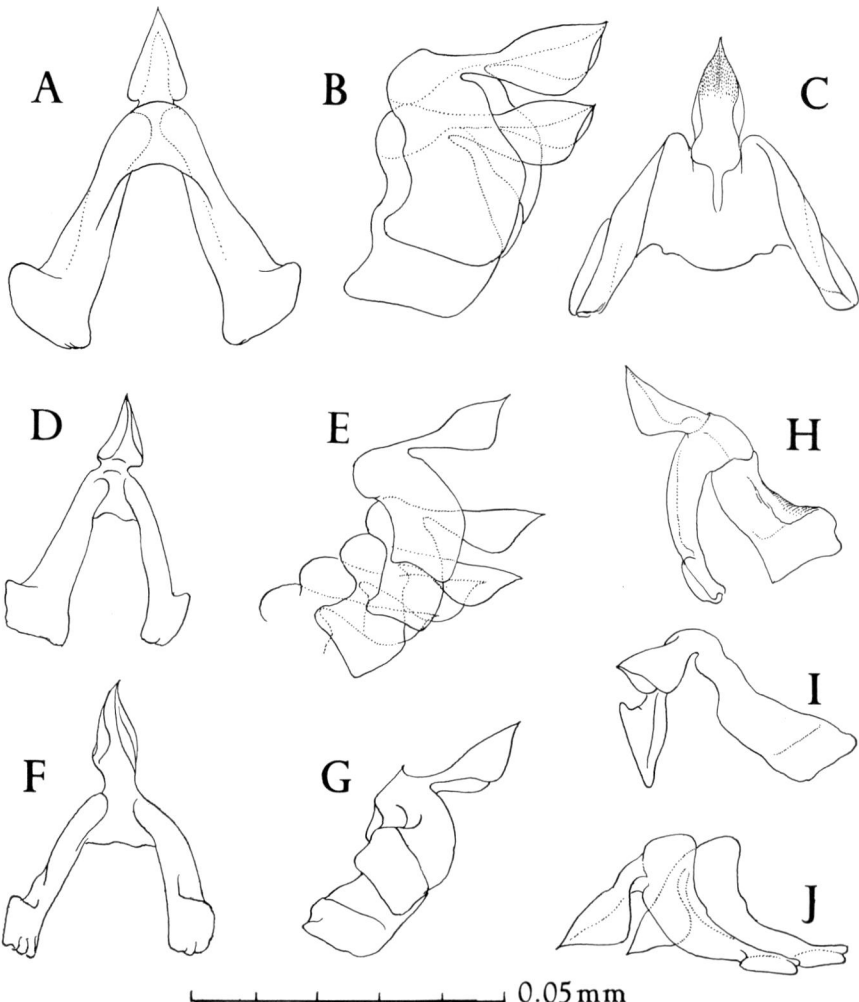

FIGURE 67. Radulae: A-C, *Volutomitra persephone*, n. sp.; D-J, *V. erebus*, n. sp.

(in addition to growth lines), especially approaching the outer lip, indicate successive pauses in growth.

No operculum. The radula is small, closely resembling that of *V. persephone*. The form of the rachidian (Fig. 67, D-J) is essentially identical, and the vestigial laterals appear as in that species.

Holotype.—USNM No. 701219, from PILLSBURY Sta. P-776.

Type-Locality.—Southern Caribbean Sea off Guajira Peninsula, Colombia (12°13.3′N, 72°50′W, 408-576 meters).

Remarks.—This species differs from *V. persephone* by its more persistent axial sculpturing, the spiral cords on the lower part of the body whorl, and the reduced fourth (anterior) columellar plication.

The material from stations P-478 and P-781 ranges in length from 19 mm to 21.5 mm, but cannot be distinguished conchologically from the larger holotype. These seem to be young individuals rather than a distinct taxon.

Class PELECYPODA

Family Dimyidae

Diagnosis.—Shell small, more or less iridescent, inequivalve, with the right valve attached, the left valve somewhat smaller and flatter; outer surface smooth or with weak radial sculpture and sometimes lamelliform concentrics; hinge margin straight, rather short, with weak outer and small inner ligament; outline rounded or ovate, often somewhat oblique; umbo scarcely projecting; mantle-line arcuate, anterior adductor muscle near the anterior margin, posterior larger, distinctly two-parted; mantle completely open, margin with papillae, without eyes; gills with uniform filaments without ascending limb; foot and labial palps rudimentary; sexes separate (Thiele, 1935: 804).

Remarks.—Moore (1970) has made some general comments about this family in his description of a new Caribbean genus and species. The Recent species now known from the western Atlantic are *Dimya argentea* Dall and *Dimyella starcki* Moore. Trawling operations aboard R/V PILLSBURY and R/V GERDA have obtained additional records of *D. argentea* and a new Caribbean species of *Dimya* possibly allied to *D. californiana* Berry from the Pacific coast of Mexico and Central America.

Deep diving by the late Dr. Thomas F. Goreau and his associates at Discovery Bay, Jamaica, and by Dr. Walter A. Starck in the Bahamas has brought to light another new dimyid quite distinct from the other Caribbean members of this family and agreeing in many respects with the genus *Dimyodon* 'Meunier-Chalmas' Fischer, a genus heretofore known as a fossil, but distinguished from it in details.

56. *Dimya argentea* Dall
Figs. 68; 71, B

Dimya argentea Dall, 1886, Bull. Mus. comp. Zool. Harv., *12*(6): 228, pl. 4, figs. 5a, 5b (extensive discussion); 1889b, Bull. U. S. natn. Mus., *37*:

FIGURE 68. Dimyidae. *Dimya argentea* Dall, Sta. P-930: upper, height 16 mm, length 15.5 mm; lower, height 15.3 mm, length 14 mm (posterior edge slightly damaged).

32, pl. 4, figs. 5a, 5b (listed only; figure identical with foregoing); 1903 (reprint of 1889).—Thiele, 1935, Handb. systematischen Weichtierkunde, 2: 804 (listed only).—Moore, 1970, J. Conch., Paris, *107*(4): 140 (listed only).

This species has been described in detail by Dall (1886). The occurrences in our faunal explorations are now reported to amplify its geographical distribution. *Dimya argentea* appears to be rather common at moderate depths in the Straits of Florida and Caribbean Sea. In our collections, it was found attached to fragments of echinoid tests and to the peripheral expansion of *Tugurium* shells, just as reported by Dall. In some cases, specimens are attached to pieces of rock or slag, and in one instance to a glass bottle. Associated with *Dimya* are numerous Foraminifera, *Sarcodictyon* (octocorallia), *Stephanoscyphus* (Scyphozoa), Bryozoa, lepadomorph and verrucomorph barnacles, and *Rhabdopleura* (Pterobranchia).

Records.—Originally reported from off Cape Hatteras, St. Croix, St. Vincent, Grenadines, and Barbados. Taken by R/V GERDA and R/V PILLSBURY at the following stations: G-179 (27°41'N, 79°11'W, 549-567 m, 1 July 1963).—G-678 (25°57'N, 78°13'W, 540-576 m, 20 July 1965).—G-720 (26°22'N, 79°11'W, 476-500 m, 3 August 1965).—G-1012 (23°35'N, 79°33'W, 508-530 m, 14 June 1968).—G-1017 (23°58'N, 79°17'W, 555 m, 15 June 1968).—P-736 (10°57'N, 65°52'W, 69-155 m, 22 July 1968).—P-848 (11°22'N, 61°26.4'W, 146 m, 2 July 1969).—P-849 (11°14.5'N, 61°46.2'W, 137-143 m, 2 July 1969).—P-876 (13° 13.9'N, 61°64.7'W, 231-258 m, 6 July 1969).—P-889 (14°04.4'N, 60° 50.8'W, 371-403 m, 7 July 1969).—P-918 (16°04.1'N, 61°25.7'W, 399-497 m, 11 July 1969).—P-930 (15°29.7'N, 61°12'W, 210-399 m, 15 July 1969).—P-984 (18°26.4'N, 63°12.6'W, 393-451 m, 22 July 1969).

57. **Dimya tigrina**, new species
Figs. 69; 71, A

Description.—Shell attached by the right valve, obliquely ovate, broader posteriorly. The hinge line is not auriculate, rather long, about 0.5 of the length of the mantle cavity (defined by the extreme limits of the pallial impression), straight, flat, transversely grooved, narrower and faintly impressed beneath the umbo. The thin, linear external ligament is visible along the whole hinge line; a deep, ovate pit beneath the umbos, closer to the posterior than to the anterior, accommodates the internal ligament. Exterior of left valve dull, pale brown marked with narrow, irregular radial streaks of darker brown; umbonal and central area externally smooth, marginal area sculptured by low, irregular radial ribs and thin, slightly raised concentric growth lamellae, resulting in a delicately scaly or frilly surface showing only faint traces of nacreous iridescence; interior of left valve

FIGURE 69. Dimyidae. *Dimya tigrina*, n. sp., holotype, Sta. P-392; height 7.85 mm (left valve) and 8.0 mm (right valve), length 9.4 mm (both valves).

porcellaneous, translucent, allowing the nacreous layer to show through marginally and showing distinct radial streaks of reddish brown; surface glossy marginally but within the pallial line dull except for the glossy muscular impressions.

Interior of attached valve cream white with conspicuous pattern of irregular radial streaks and spots of reddish brown most intense at the margins; surface marginally glossy, centrally dull except for the glossy muscular impressions; a wide band of irregular, narrow radial grooves and wrinkles lies within the pallial line, narrowing toward the hinge where the wrinkles break up into granulations that merge with the cross-striated hinge line.

Anterior muscular impression elliptical, near the end of the hinge line and closer to the umbo than is the posterior impression. Posterior muscular impression well removed from the end of the hinge line, distinctly double, that in the left valve more clearly bilobed than that in the right.

The wide margin of the right valve is free of the substrate and flares broadly; its outer surface is marked with low, broad radials crossed by scaly concentrics, resulting in an imbricated appearance.

Measurements.—Left valve, length 9.4 mm, height at umbo, 7.85 mm; right valve, length 9.4 mm, height at umbo 8.0 mm (holotype).

Holotype.—USNM No. 701220, from PILLSBURY Sta. P-392.

Type-Locality.—PILLSBURY Sta. P-392, off Punta Piedras, Colombia: 9°45.1'N, 76°09.1'W, 79-75 meters; 16 July 1966.

Remarks.—This species is notable for its conspicuous color pattern of radial brown streaks and spots, and its minimal iridescence. The left valve is only slightly smaller than the right. *Dimya tigrina* seems to have a general resemblance to the eastern Pacific *Dimya californiana* Berry, in which the left valve is "irregularly clouded with Sayal brown," and in which the area outside the pallial line is "beautifully silvery-pearly" (Berry, 1936: 126-127), but the type is in a private collection, and no authentic material has been available for comparison. *Dimya radiata* Kuroda, from Japanese waters, also is marked with radiating brown streaks.

Although Berry noted that the adductor impressions of *D. californiana* are proportionately larger than in other species, there are few reliable taxonomic characters whereby the species of *Dimya* can be distinguished. A thorough revision of the genus on a worldwide basis is needed to evaluate morphological characters of the shells and to place the taxonomy of these extremely interesting pelecypods on a firm basis.

Basiliomya, new genus

Diagnosis.—Shell subcircular in outline, translucent, with negligible iridescence; left valve smooth externally, but sometimes reproducing irregularities of the substrate; radial sculpture on outer surface of free margin of right valve; edge of right valve more or less widely extended as a thin, lobate marginal frill. Hinge of right valve with a blunt, triangular tooth on each side of the internal ligament and a shallow socket at each end of the hinge line; left valve with a shallow groove on each side of the internal ligament and a blunt tooth at each end of the hinge line; a series of interlocking small teeth and pits around the perimeter of both valves. Anterior adductor

FIGURE 70. Dimyidae. *Basiliomya goreaui*, n. sp., from Discovery Bay, Jamaica. Upper, lateral and oblique views of holotype; lower (left to right), inner and outer views of left valve of holotype; paratype, complete specimen in situ.

near end of hinge line; posterior adductor remote from hinge, conspicuously bilobed; pallial impression marked by a row of shallow pits.

Type-Species.—*Basiliomya goreaui*, new species, here designated.

Gender.—Feminine.

Remarks.—This distinctive genus resembles the fossil *Dimyodon* Meunier Chalmas (*in* Fischer, 1886: 937) in the characters of the hinge, but has an additional tooth at each end of the hinge line in the left valve and corresponding sockets in the right. It differs further in having the posterior adductor impression distinctly bilobed as in *Dimya*, whereas it is simple in *Dimyodon* as illustrated by Fischer (1886).

58. **Basiliomya goreaui**, new species
Figs. 70; 71, C-E

Description.—Shell subcircular, attached by the right valve, translucent porcellaneous white with faint pearly reflections internally, especially in the left valve. The right valve has a more or less strongly up-turned ventral margin rising vertically from the substrate, hence the valve is deep, whereas the left valve is almost flat, with only a weak concavity within the pallial line. Externally, the left valve is unsculptured (although it may show irregularities due to the surface of the substrate), convex in the umbonal area and concave marginally, recessed within the right valve and fitting it tightly. Margin of the right valve expanded beyond the edge of the left valve to form a thin flange with several blunt marginal digitations; ventrally this flange is bent toward the substrate so the marginal digitations are directed downward, but anteriorly and posteriorly it extends as two broad, fluted lobes that may be either spread outward and downward, or reflected upward and curled over the ends of the free valve. The ventral junction of the right valve with the substrate is strengthened by a fluted, buttresslike structure that evidently was the first marginal expansion formed when the shell reached its definitive diameter; subsequent shell-growth proceeds vertically with respect to the substrate, and although a series of marginal flanges may be formed (Fig. 71, E), most of the specimens observed have only one. The width and elegance of the frill varies, presumably according to local conditions, but it is present in all specimens examined from Jamaica and the Bahamas.

The hinge line is straight, not auriculate, about 0.42 of the length of the upper valve; in the lower (right) valve, the ligament is flanked on both sides by a low, bluntly triangular tooth, and there is a shallow, rounded pit at each end of the hinge line. The dorsal aspect of the two teeth is transversely striated where it articulates with triangular depressions, correspondingly striated, on either side of the ligament in the upper valve. In the upper valve, a low, bluntly rounded tooth stands at each end of the hinge

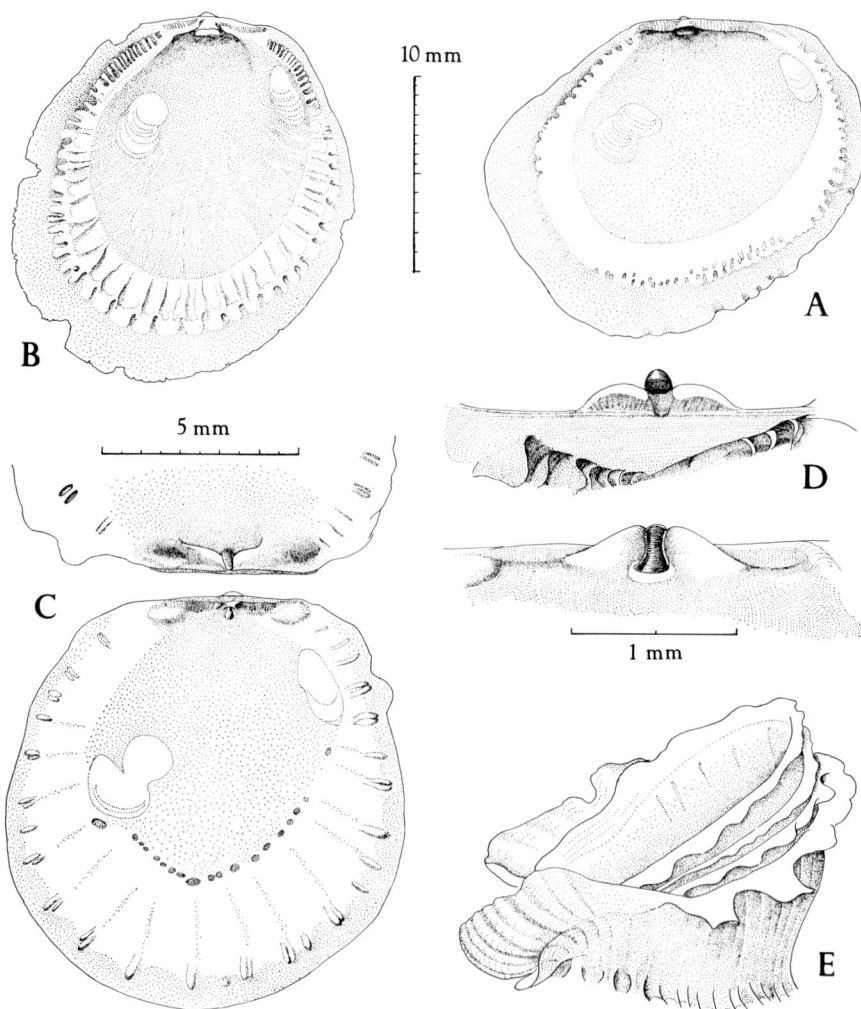

FIGURE 71. Dimyidae. A, *Dimya tigrina*, n. sp., holotype; interior of left valve; B, *D. argentea* Dall, Sta. P-876; C, *Basiliomya goreaui*, n. sp., paratype; interior of left valve and umbonal area of right, showing general features of hinge; D, *Basiliomya goreaui*, n. sp., paratype; hinge of right valve, dorsal and ventral aspects; E, *Basiliomya goreaui*, n. sp., paratype; oblique view of right valve. (The 10-mm scale applies to A and B; 5-mm scale to C and E; 1-mm scale to D.)

line and articulates with the sockets in the lower valve. No external ligament can be seen and, if present, must be extremely delicate. The internal ligament is attached in the umbonal depression of the right valve and in a subumbonal pit in the left valve.

Within the margin of the left valve there is a series of radially oriented, elongated denticles, sometimes simple, sometimes paired, which fit within correspondingly simple or paired pits in the opposite valve (Fig. 71, C). Interior of both valves glossy, but roughened within the pallial impression of the left valve; inner boundary of pallial impression marked by a series of conspicuous round or ovate shallow pits; in the right valve the upward curvature of the shell begins at the pallial impression and the pits are difficult to observe. Anterior muscular impression elliptical, removed by less than its own length from the end of the hinge line. Posterior muscular impression large, bilobed, situated about midway between dorsal and ventral margins.

Measurements.—Length of left valve, 4.7 mm, height 4.8 mm; length (including marginal frill) of right valve, 7.1 mm, height 5.9 mm (holotype). Length of left valve 4.5 mm, height 4.4 mm; length of right valve (posterior frill damaged) 7.0 mm, height 5.7 mm (illustrated paratype, Fig. 70).

Holotype.—USNM No. 701221; Discovery Bay, Jamaica, depth 170 feet; attached to dead branch of scleractinian coral (*Madracis*). Collected by Dr. Thomas F. Goreau, December 25, 1964.

Paratypes.—Discovery Bay, Jamaica, West Bull, depth 200 feet; attached to lower surface of scleractinian coral (*Agaricia*). Five right valves, collected by J. B. C. Jackson, 1970.

Discovery Bay, Jamaica, 175 feet; attached to lower surface of scleractinian coral (*Agaricia*). Five complete specimens and two right valves, collected by J. B. C. Jackson, October 1970.

Reef off Goat Bay, Fresh Creek, Andros I., Bahamas, 75-100 feet; attached to bottom of large coral (*Agaricia*). Complete specimens and empty right valves, collected by J. D. Starck and Peter Hopper, 19 December 1969.

Andros Island, Bahamas; attached to scleractinian coral (*Scolymia*). One complete specimen and one right valve, collected by W. A. Starck and J. D. Starck, April 20, 1970.

Family Spondylidae
59. *Spondylus gussoni* Costa
Fig. 72

Spondylus gussoni O. G. Costa, 1829, Cat. Sist.: xlii.—Philippi, 1836, Enum. Moll. Siciliae, *1*: 87, pl. 5, fig. 16.—Dall, 1886, Bull. Mus. comp. Zool. Harv., *12*: 227.—Locard, 1898, Expéd. Travailleur & Talisman, *4*(2): 420, pl. 18, figs. 1-8.

FIGURE 72. Spondylidae. *Spondylus gussoni* O. G. Costa, Sta. P-887.

Not *Spondylus gussoni*, Webb, 1935, Handbook: 54 (= *S. regius* L.?).—
Barrett & Patterson, 1970, Shells and Shelling: 54, fig. 1 (= *S. ictericus* Rve.)

Records.—PILLSBURY Sta. P-887, off St. Lucia, 14°10.6′N, 60°55.8′W, 73-110 m, 7 July 1969; one living specimen.—P-1303, Caribbean Sea, 20 miles ESE of Sto. Domingo, island of Hispaniola, 18°21.0′N, 69°14.3′W, depth 174 m, 21 July 1970. Single attached valve.

Remarks.—Taken by the BLAKE off Nevis, Barbados, Grenada, and Yucatan, at depths from 168 to 686 meters. Dall stated, "The specimens have been compared with authentic European examples, and agree precisely." Compared with the PILLSBURY specimens, those from 207 fm (= 379 m) off Morocco in the U. S. National Museum of Natural History (196472, Jeffreys Coll.) are similar in size and shape but thicker in construction, with about 50 equal ribs without spines (wear?). The specimen collected by the TRAVAILLEUR in the Bay of Biscay (USNM 62240, Jeffreys Coll.) has the anterior auricle of the left valve more prominent, ribs equal but more numerous (about 75), with close-set sharp, short, decurved thorns. Although nothing is known of the variation in true *S. gussoni*, it seems possible that eastern Atlantic material pertains to more than one species and quite probable that western Atlantic material is specifically distinct.

The name *S. gussoni* has been attached to the common shallow-water *S. ictericus* in some popular literature and not infrequently is applied in that sense to specimens by collectors. There is clearly no basis for this usage.

SUMARIO

MOLUSCOS NUEVOS Y RAROS COLECTADOS POR LOS BARCOS DE INVESTIGACIONES JOHN ELLIOTT PILLSBURY Y GERDA EN EL ATLÁNTICO OCCIDENTAL TROPICAL

Se reportan e ilustran 59 especies nuevas o raras de moluscos marinos, 55 gasterópodos y cuatro pelecípodos, del área del Caribe. Entre éstos se describen las siguientes nuevas taxa: *Calliostoma olssoni*, n. sp., *Thelyssa callisto*, n. gen., n. sp., *Lischkeia deichmannae*, n. sp. (Trochidae); *Sconsia nephele*, n. sp. (Cassididae); *Typhis* (*Siphonochelus*) *tityrus*, n. sp. (Muricidae); *Columbarium* (*Peristarium*) *electra*, n. subgen., n. sp., *C.* (*P.*) *merope*, n. sp., *C.* (*P.*) *aurora*, n. sp. (Columbariidae); *Coralliophila fax*, n. sp., *C. sentix*, n. sp. (Coralliophilidae); *Teramachia chaunax*, n. sp. (Turbinellidae); *Lyria* (*Cordilyria*) *cordis*, n. subgen., n. sp., *Scaphella* (*Clenchina*) *evelina*, n. sp., *Volutomitra persephone*, n. sp., *V. erebus*, n. sp. (Volutidae); *Dimya tigrina*, n. sp., *Basiliomya goreaui*, n. gen., n. sp. (Dimyidae). Se da el primer reporte de ejemplares vivos de *Bathygalea coronadoi* (Crosse). *Mesorhytis meekiana* Dall, originalmente asignada a la familia Fasciolariidae, es transferida al género *Teramachia* y colocada en la familia Turbinellidae. Las conchas son ilustradas por medio de fotografías, las rádulas y los opérculos de algunas de las especies son ilustradas con dibujos y el conjunto de la anatomía de la cavidad del manto es ilustrada en *Oocorys sulcata* Fischer, *Vasum capitellum* (Linné) y *Lyria cordis*, n. sp.

LITERATURE CITED

ABBOTT, ROBERT TUCKER
    1958. The marine mollusks of Grand Cayman Island, British West Indies. Monogr. Acad. nat. Sci. Philad., No. 11: [viii] + 138 pp. + 5 pls. [+ 16 pp. index in reprint ed., 1967].

BARRETT, RALPH AND DAVID PATTERSON (EDS.)
1970. Shells and shelling. New edition. Fletcher Pub. Co., 64 pp., ill. (Also earlier eds.)

BERRY, S. STILLMAN
1936. A new *Dimya* from California. Proc. malac. Soc. Lond., *22*(3): 126-128, pl. 13B.

BOSS, KENNETH J.
1967. *Thyasira disjuncta* (Gabb, 1866) in the Caribbean Sea. Bull. Mar. Sci., *17*(2): 386–388.

BULLIS, HARVEY R., JR.
1964. Muricidae (Gastropoda) from the northeast coast of South America, with descriptions of four new species. Tulane Stud. Zool., *11*(4): 99-107, figs. 1-8.

BURGESS, C. M.
1970. The living cowries. A. S. Barnes & Co., South Brunswick, New York, 389 pp., 49 pls. paginated in the text.

CLENCH, WILLIAM J.
1944. The genus *Columbarium* in the western Atlantic. Johnsonia, *1*(15): 1-4.
1946. The genera *Bathyaurinia*, *Rehderia* and *Scaphella* in the western Atlantic. Johnsonia, *2*(22): 41-60, 31 "plates" in the text.
1953. The genera *Scaphella* and *Auriniopsis* in the western Atlantic. Johnsonia, *2*(32): 376-380, pls. 186-187.
1959. The genus *Columbarium* in the western Atlantic. Johnsonia, *3*(39): 330-331, pl. 173.

CLENCH, WILLIAM J. AND ROBERT TUCKER ABBOTT
1943. The genera *Cypraecassis*, *Morum*, *Sconsia* and *Dalium* in the western Atlantic. Johnsonia, *1*(9): 1-8, "plates" 1-4 in the text.

CLENCH, WILLIAM J. AND C. G. AGUAYO
1940. Notes and descriptions of new deep-water Mollusca obtained by the Harvard-Habana Expedition off the coast of Cuba. III. Mem. Soc. Cub. Hist. nat., *14*(1): 77-94, pls. 14-16.
1941. Notes and descriptions of new deep-water Mollusca obtained by the Harvard-Havana Expedition off the coast of Cuba. IV. Mem. Soc. Cub. Hist. nat., *15*(2): 177-180, pl. 14.

CLENCH, WILLIAM J. AND RUTH TURNER
1952. The genera *Epitonium* (part II), *Depressiscala*, *Cylindriscala*, *Nystiella* and *Solutiscala* in the western Atlantic. Johnsonia, *2*(31): 289-356, "plates" 131-177 in the text.
1964. The subfamilies Volutinae, Zidoninae, Odontocymbiolinae and Calliotectinae in the western Atlantic. Johnsonia, *4*(43): 129-180, pls. 80-114.
1967. A new species of *Lyria* (Volutidae) from Hispaniola. Nautilus, *80*(3): 83-84, figs. 1-3.
1970. The family Volutidae in the western Atlantic. Johnsonia, *4*(48): 369–372, pls. 172–174.

COOKE, A. H.
1922. The radula of the Volutidae. Proc. malac. Soc. Lond., *15*: 6-12, figs. 1-8.

COOMANS, H. E.
1963. Systematics and distribution of *Siphocypraea mus* and *Propustularia*

surinamensis (Gastropoda, Cypraeidae). Stud. Fauna Curaçao, *15*: 51-71.

COTTON, BERNARD C.
1959. South Australian Mollusca Archaeogastropoda. Handbook of the flora and fauna of south Australia. Gov't. Printer, Adelaide, 449 pp., 215 figs., 1 pl.

DALL, WILLIAM HEALEY
1881. Preliminary report on the Mollusca. Reports on the results of dredging . . . in the Gulf of Mexico, and in the Caribbean Sea, 1877-79, by the United States Coast Survey steamer "Blake" . . . . Bull. Mus. comp. Zool. Harv., *9*(2): 33-144.

1886. Report on the Mollusca. Part I. Brachiopoda and Pelecypoda. Reports on the results of dredging . . . in the Gulf of Mexico (1877-78) and in the Caribbean Sea (1879-80), by the U. S. Coast Survey steamer "Blake" . . . . Bull. Mus. comp. Zool. Harv., *12*(6): 171-318, pls. 1-9.

1889a. Report on the Mollusca. Part II.—Gastropoda and Scaphopoda. Reports on the results of dredging . . . in the Gulf of Mexico (1877-78) and in the Caribbean Sea (1879-80), by the U. S. Coast Survey steamer "Blake" . . . . Bull. Mus. comp. Zool. Harv., *18*: 1-492, pls. 10-40.

1908. The Mollusca and the Brachiopoda. Reports on the dredging operations off the west coast of Central America . . . by the U. S. Fish Commission steamer "Albatross," during 1891 . . . . XXXVIII. Reports on the scientific results of the expedition to the eastern tropical Pacific . . . by the . . . "Albatross," from October, 1904, to March, 1905 . . . . Bull. Mus. comp. Zool. Harv., *43*(6): 205-487, pls. 1-22.

1921. Summary of the marine shellbearing mollusks of the northwest coast of America, from San Diego, California, to the Polar Sea, mostly contained in the collection of the United States National Museum, with illustrations of hitherto unfigured species. Bull. U. S. natn. Mus., *112*: 217 pp., 22 pls.

DANCE, S. PETER
1966. Shell collecting. An illustrated history. University of California Press, Berkeley and Los Angeles, 344 pp., 31 figs., 35 pls.

1969. Rare Shells. University of California Press, Berkeley and Los Angeles, 128 pp., 24 pls.

DANCE, S. PETER AND WILLIAM KEITH EMERSON
1967. Notes on *Morum dennisoni* (Reeve) and related species (Gastropoda: Tonnacea). Veliger, *10*(2): 91-98, pl. 12.

DARRAGH, THOMAS A.
1969. A revision of the family Columbariidae (Mollusca: Gastropoda). Proc. R. Soc. Vict., *83*(1): 63-119, figs. 1-24, pls. 2-6.

DESHAYES, G.-P.
1856. Description d'espèces nouvelles. J. Conch., Paris, *5*: 78-82, pl. 3.

EMERSON, WILLIAM KEITH
1967. A new species of *Morum* from Brazil, with remarks on related species. (Gastropoda: Tonnacea). Veliger, *9*(3): 289-292, pl. 39.

EMERSON, WILLIAM KEITH AND ANTHONY D'ATTILIO
1963. A new species of *Latiaxis* from the western Atlantic. Am. Mus. Novit., No. 2149: 1-9, figs. 1-6.

EMERSON, WILLIAM KEITH AND WILLIAM E. OLD, JR.
1965. New records for *Cypraea surinamensis*. Nautilus, 79(1): 26-30, pl. 3, figs. 1-2.
1966. *Cypraea* (*Propustularia*) *surinamensis* Perry from Brazil. Nautilus, 80(2): 70-71.

FISCHER, PAUL
1880- Manuel de conchyliologie et de paléontologie conchyliologique ou
1887 historie naturelle des mollusques vivants et fossiles. F. Savy, Paris, xxiv + 1369 pp., 23 pls.

GERTMAN, RICHARD L.
1969. Cenozoic Typhinae (Mollusca: Gastropoda) of the western Atlantic region. Tulane Stud. Geol., 7(4): 143-191, figs. 1-3, pls. 1-8.

GRAY, JOHN EDWARD
1857. Guide to the systematic distribution of Mollusca in the British Museum. Part 1. London, xii + 230 pp.

KEEN, A. MYRA
1958. Sea shells of tropical west America. 624 pp., 11 pls., and 606, 12, 1037, 3, and 51 text-figs.

KIENER, L. C.
1835. Genre Pourpre (Purpura, Lam.). Spécies général et iconographie des coquilles vivantes. Pp. 1-151, pls. 1-46.

MATTHEWS, HENRY RAMOS
1967. Notas sôbre os cipreídos do nordeste Brasileiro. Arq. Estac. Biol. Mar. Univ. Ceará, 7(1): 15-18, figs. 1-8.

MEEK, F. B.
1876. A report on the invertebrate Cretaceous and Tertiary fossils of the Upper Missouri Country. Report of the United States Geological Survey of the Territories, 9, lxiv + 629 pp., 85 text-figs., 43 pls.

MOORE, DONALD R.
1970. A new genus and species of Dimyidae from the Caribbean coast of Mexico. J. Conch., Paris, 107(4): 137-141 + 1 pl.

MOREIRA LEME, JOSÉ LUIZ AND LÍCIA PENNA
1969. Occorrência de *Mikadotrochus* no Brasil, com descrição de uma nova espécie (Gastropoda, Pleurotomariidae). Papéis Dep. Zool. S Paulo, 22(21): 225-230, figs. 1, 2.

OLSSON, AXEL A.
1964. Neogene mollusks from northwestern Ecuador. Paleontological Research Institution, Ithaca, N. Y., 258 pp., 38 pls.
1965. A review of the genus *Voluta* and the description of a new species. Bull. Am. Paleont., 49(224): 653-672, 4 pls.

PEILE, A. J.
1922. Some notes on radulae. Proc. malac. Soc. Lond., 15: 13-18, figs. 1-5.

PERRY, GEORGE
1811. Conchology, or the natural history of shells: containing a new arrangement of the genera and species . . . . London. 61 pls. + descriptive letterpress.

PHILIPPI, R. A.
1836- Enumeratio molluscorum siciliae. Vol. 1, Berolim, 1836, pp. I-XIV,
1844 1-268, pls. I-XII. Vol. 2, Halis Saxonum, 1844, pp. I-IV, 1-303, pls. 13-28.

PILSBRY, HENRY A. AND AXEL A. OLSSON
   1953. Materials for a revision of east coast and Floridian volutes. Nautilus, 67(1): 1-13, pls. 1-3.
   1954. Systems of the Volutidae. Bull. Am. Paleont., 35(152): 1-36, pls. 1-4 (25-28).
REEVE, LOVELL AUGUSTUS
   1857. Description of seven new shells from the collection of the Hon. Sir David Barclay of Port Louis, Mauritius. Proc. zool. Soc. Lond., Part 25: 207-210, pls. 37-38.
ROBERTSON, ROBERT
   1957. The subgenus *Halopsephus* Rehder, with notes on the western Atlantic species of *Turbo* and the subfamily Bothropomatinae Thiele. J. Wash. Acad. Sci., 47(9): 316-319, figs. 1-3.
SHIKAMA, TOKIO AND MASUOKI HORIKOSCHI
   1963. Genshoku Zukan Sekai no Kai. Selected shells of the world illustrated in colours. [In Japanese.] Hokuryukan, Tokyo [x] + 102 pls. + 154 pp.
SMITH, MAXWELL
   1939. An illustrated catalog of the Recent species of the rock shells. Muricidae, Thaisidae and Coralliophilidae. Tropical Laboratory, Lantana, Florida, 83 pp., 21 pls.
THIELE, JOHANNES
   1929- Handbuch der systematischen Weichtierkunde. Jena, Verlag von
   1935 Gustav Fischer. Theil 1, pp. 1-376, 1929; Theil 2, pp. 377-778, 1931; Theil 3, pp. vi + 779-1022, 1934; Theil 4, pp. 1023-1154, 1935.
TRYON, GEORGE WASHINGTON
   1880. Manual of conchology; structural and systematic. Vol. 2. Muricinae, Purpurinae. Philadelphia, published by the author. 289 pp., 70 pls.
   1882. Manual of conchology; structural and systematic. Vol. 4. Nassidae, Turbinellidae, Volutidae, Mitridae. Philadelphia, published by the author. 276 pp., 58 pls.
VERRILL, ADDISON EMERY
   1885. Third catalogue of Mollusca recently added to the fauna of the New England coast . . . . Trans. Conn. Acad. Arts Sci., 6: 395-452, pls. 42-44.
VOKES, EMILY H.
   1963. Cenozoic Muricidae of the western Atlantic region. Part 1—*Murex* sensu stricto. Tulane Stud. Geol., 1(3): 93-123, pls. 1-4.
   1964. Supraspecific groups in the subfamilies Muricinae and Tritonaliinae (Gastropoda: Muricidae). Malacologia, 2(1): 1-41, pls. 1-3.
   1967. Observations on *Murex messorius* and *Murex tryoni*, with the description of two new species of *Murex*. Tulane Stud. Geol., 5(2): 81-90, figs.
WATSON, ROBERT BOOG
   1879. Mollusca of H. M. S. 'Challenger' Expedition. III. Trochidae, viz. the genera *Seguenzia, Basilissa, Gaza,* and *Bembix*. J. Linn. Soc. Lond., Zool., 15: 586-605.
   1886. Report on the Scaphopoda and Gasteropoda collected by H. M. S. CHALLENGER during the years 1873-76. Rep. Sci. Res. Challenger, Zool., 15(Part 42): 1-680, pls. 1-50.

WEAVER, CLIFTON STOKES AND JOHN ELEUTHÈRE DU PONT
    1970.   Living volutes. A monograph of the Recent Volutidae of the world. Delaware Museum of Natural History (Greenville, Del.) Monograph Series No. 1, xv + 375 pp., 79 pls.

WEBB, WALTER FREEMAN
    1935.   A Handbook for shell collectors. Second ed. Rochester, N. Y. & St. Petersburg, Fla., 146pp. (Also later editions.)

WOODRING, WENDELL P. AND AXEL A. OLSSON
    1957.   *Bathygalea*, a genus of moderately deep-water and deep-water Miocene to Recent cassids. Prof. Pap., U. S. Geol. Surv., *314B*: 21-26, pls. 7-10.

WORK, ROBERT C.
    1969.   Systematics, ecology, and distribution of the mollusks of Los Roques, Venezuela. Bull. Mar. Sci., *19*(3): 614-711, figs. 1-4.